SEVEN STAGES OF RECONSTRUCTING THE FACE OF AN EXTINCT HUMAN ANCESTOR.

[PAGE 1]
Stage I: Cast of the Sima 5 *Homo heidelbergensis* skull originally found in July 1992 by a team led by Juan-Luis Arsuaga in the Sima de los Huesos (Pit of Bones), Atapuerca, Spain.

[PAGE 2]
Stage II: The Sima 5 skull with missing or damaged, teeth and eye orbit reconstructed.

[PAGE 3]
Stage III: Some deep soft tissues are applied to the reconstructed Sima 5 skull.

[PAGE 4]
Stage IV: Most of the superficial soft tissues are applied to the Sima 5 reconstruction.

[PAGE 5]
Stage V: Right half of a complete deep and superficial soft tissue reconstruction; and left half a complete facial reconstruction of the Sima 5 man.

[PAGE 6]
Stage VI: Skin has been applied to soft tissue and skull.

[PAGE 7]
Stage VII: To complete the reconstruction, facial hair is added, revealing the face of a 350,000-year-old *Homo heidelbergensis* man from what is now Atapuerca, Spain.

[FACING PAGE] James Howard McGregor (1872-1955) created this classic series of sculpture busts, depicting Piltdown Man (*Eoanthropus dawsonii*, later revealed as a fossil forgery), Java Man (*Pithecanthropus erectus*, now ***Homo erectus***), Neanderthal Man, and modern or Cro-Magnon Man.

DEDICATED TO THE MEMORY OF JAMES HOWARD MCGREGOR
FOR WITHOUT HIS INSPIRATION THIS BOOK WOULD NOT EXIST.

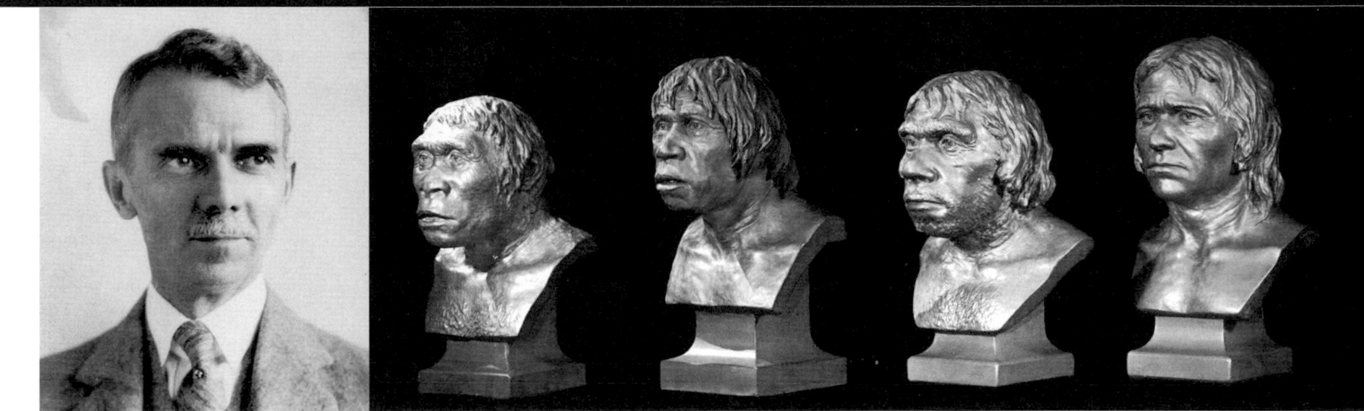

THE LAST HUMAN

A GUIDE TO TWENTY-TWO SPECIES OF EXTINCT HUMANS

CREATED BY G.J. SAWYER
AND VIKTOR DEAK

TEXT BY
ESTEBAN SARMIENTO
G.J. SAWYER
RICHARD MILNER

WITH CONTRIBUTIONS BY
DONALD C. JOHANSON
MEAVE LEAKEY and
IAN TATTERSALL

A PETER N. NÈVRAUMONT BOOK

Yale University Press
New Haven and London

All photographs, except as noted, by Viktor Deak.
Reconstructed skulls of *Australopithecus afarensis*, *Australopithecus africanus* (adult), *Australopithecus africanus* (child), *Homo antecessor*, *Homo ergaster*, *Homo floresiensis*, *Homo georgicus*, *Homo habilis*, *Homo heidelbergensis*, *Homo neanderthalensis*, *Homo pekinensis*, *Homo rhodesiensis*, *Homo sapiens* (Qafzeh), *Paranthropus aethiopicus*, *Paranthropus boisei*, and *Paranthropus robustus* by G.J. Sawyer and Viktor Deak.
Reconstructed Neanderthal skeleton by G.J. Sawyer and Blaine Maley.
Model heads of *Australopithecus afarensis*, *Australopithecus africanus*, *Homo ergaster*, *Homo floresiensis*, *Homo Heidelbergensis*, *Homo habilis*, *Homo neanderthalensis*, *Homo pekinensis*, *Homo rudolfensis*, *Paranthropus boisei*, and *Paranthropus robustus* by G.J. Sawyer and Viktor Deak.

Jacket and book design by Cathleen Elliott.
Maps by Kale Halprin.

Printed in China by Everbest Printing Company through Four Colour Imports.

Library of Congress Control Number: 2006928034

ISBN-13: 978-0-300-10047-1 (cloth : alk. paper)
ISBN-10: 0-300-10047-7 (cloth : alk. paper)

A catalogue record for this book is available from the British Library.

The paper in this book meets the guidelines for permanence and durability of the Committee on Production Guidelines for Book Longevity of the Council on Library Resources.

10 9 8 7 6 5 4 3 2 1

Produced by Nèvraumont Publishing Company, New York City.

CONSULTANTS

Peter Brown
Professor
Department of Archaeology and Palaeoanthropology
School of Human & Environmental Studies
University of New England
Armidale, Australia

Robert Franciscus
Associate Professor
Department of Anthropology
University of Iowa

Eliot Goldfinger
Artist/Anatomist

Charles Hilton
Assistant Professor
Department of Anthropology
Western Michigan University

Trent Holliday
Associate Professor
Department of Anthropology
Tulane University

Ralph Holloway
Professor
Department of Anthropology
Columbia University

Samuel Marquez
Director of Gross Anatomy
Research Assistant Professor
SUNY Downstate Medical Center College of Medicine

Jay Matternes
Paleoartist

Fred Spoor
Professor of Evolutionary Anatomy
University College London

Phillip V. Tobias
Honorary Professor of Palaeoanthropology
Bernard Price Institute for Palaeontological Research
Honorary Professor in Zoology
Professor and Head
Department of Anatomy
University of Witwatersrand

CONTENTS

TWENTY-TWO SPECIES OF EXTINCT HUMAN ANCESTORS

by Esteban Sarmiento | 24

FOREWORD

For a fossil hunter like myself there are few experiences that make the heart beat faster than the discovery of one of our ancestor's skulls. I recall with great clarity when my colleague, Yoel Rak, announced his discovery of a 3.0 million-year-old male skull of *Australopithecus afarensis* during our 1992 Hadar Research Project in the Afar region of Ethiopia. Staring into the fossilized, bony outline of a face, long dead and long forgotten, is a rare privilege, one that rewards like few others. This fossil was once a living individual, once a member of a larger group of individuals, all of whom have now vanished forever, but because of the vagaries of geological serendipity, this one individual sees the light of day once more.

During earlier expeditions to Hadar in the nineteen seventies, great excitement was generated by my discovery of the partial skeleton now popularly known as "Lucy." While Lucy was remarkably complete, with parts of arms, legs, backbone, and pelvis, the skeleton preserved only the jaw bone, not the face or braincase—her skull was never recovered. After arduous years of searching the sediments at Hadar, we continued to come up empty-handed. A skull continued to elude us. The 1992 skull find at Hadar was a landmark, and at long last a critical specimen, which would go far in our goal towards filling out the anatomy of *Australopithecus afarensis*, had finally come to light.

Why are skulls so vital in our assessment of fossil species, in this particular case, our own ancestors? A majority of anatomical features that define a species are present in the teeth and skull and are essential for a comprehensive understanding of an extinct species. It really is no different for modern humans. We seek out the familiar and definitive features in the faces of our friends to make an identification. It is our physiognomy, not the shape of our legs or arms, that we use when describing one another. To form an idea of what an *Australopithecus afarensis* skull would look like previous to the 1992 discovery, a composite reconstruction of a skull was carefully assembled incorporating fossil fragments from twelve adult specimens. We surmised this was a pretty good approximation and could at least stand as a "hypothesis" of what we hoped would eventually be found. Yoel's skull would provide the test for that hypothesis.

All through the meticulous and time-consuming process of refitting the broken fragments of the 1992 skull from Hadar, officially catalogued as A.L. 444-2, my imagination would frequently spiral off in the direction of trying to envision what this creature must have looked like in the flesh. The skull itself was massive, in fact, the largest known *Australopithecus* skull ever found, and I knew he would have possessed an intimidating visage.

Since *Australopithecus afarensis* has a significant level of sexual dimorphism, much like what we see in other species of *Australopithecus*, we knew that this skull would have been different from that of a female like Lucy. As a result of continuing strategic field research at Hadar, a female skull,

BY **DONALD C. JOHANSON**

A.L. 822-1, was found in 2002 eroding from a 3.1 million-year-old geological layer. This specimen, beautifully preserved with the oldest most complete *Australopithecus* mandible thus far found, is currently under study and promises to provide a long anticipated view of a female *A. afarensis* skull, albeit a little larger than the one Lucy would have carried atop her diminutive body.

Innumerable thoughts come to mind when we peer into the splendid reconstructed faces of our long-vanished ancestors presented in this book. If we were able to travel back into time and stand alongside these creatures as living and breathing individuals, I wonder at which stage we would be able to communicate with each other? As affectionate a name as "Lucy" is, I have the sense that a troop of *Australopithecus afarensis* individuals would have fled from me, much like unhabituated apes do in the wild. If I managed to get within closer range, would they charge and attack me? How many species would look at me as an oddity, something upright walking like themselves, but so different looking? Might my encounters with more recent species, such as *Homo heidelbergensis* and Neanderthals, elicit fear and aggression because I would be viewed as a potential competitor?

These and so many other questions will go unanswered, but at least this gem of a book with such provocative reconstructions of our predecessors will stimulate us all to think more deeply about who we are and how we fit into the broader scheme of human evolution. We are reminded of our humble origins and how unique and treacherous the evolutionary path must have been from the most ancient ancestors to ourselves, the only living descendent, who remarkably enough has the capacity to ponder its past.

WE WERE

INTRODUCTION

One of the remarkable characteristics of human beings is their intense intellectual curiosity, not least about their own origins. All of us want to know where we came from, both as individuals and as a species; and every society that has ever been documented has invented its own origin stories and myths to fill this deep-seated need. This book is about the scientific story, which turns out not to be the story that we would perhaps have instinctively expected. Why? Well, because *Homo sapiens* is the only species of its general kind that exists today—in a very profound sense we are alone in the world—and since this is the familiar situation for us, we naturally take it to be the "normal" condition for a species such as ours. As a result, we tend to think of ourselves as the end product (at least for the time being) of a single lineage that was gradually improved, generation by generation, by natural selection.

For substantiation of this view, we often appeal to the matter of brain size, always a major issue to a species whose huge brain is among the most obvious of its unique attributes. Three million years ago our precursors had brains that were little, if any, larger than those of apes of similar body size. Two million years later those brains had doubled in their relative size. And today our brains have doubled again, seemingly proof that our biological history has been guided to its current exalted status by the gentle perfecting hand of natural selection.

If this indeed is what we thought, we could hardly have been more wrong. Gary Sawyer, Viktor Deak, Esteban Sarmiento, and their collaborators have looked at the fossil record in a different way, and have evoked a very different picture of our evolution. This alternative view sees human prehistory as the story of a diverse hominid family that sprawled rather than snaked across the ages. From the beginning this story was a dynamic one, a drama of numerous species that were spawned in various localities and that then set forth to do battle in the ecological arena. These struggles for ecological space were fought with relatives both close and remote, and they took place within an environmental theater whose aspect was constantly changing. What's more, the stakes were high: the victors in these battles gave rise to descendant species, while the rest became extinct without issue.

Still, from the very beginning the stage was almost always well populated with actors. For throughout the history of the hominid family (the taxonomic group that contains us and our close relatives), the world was typically home to several different kinds of pre-human or human. At least from time to time and from place to place, a number of different hominid species could even be found on the very same landscape. Up until quite recently, there seems to have been one or more coexisting hominid species, sometimes sharing the same landscape. It is only today that we find a single kind of hominid in the world. The most important message encoded in this fact is that there is something very special and unusual about *Homo sapiens* that makes it a uniquely dangerous competitor. But this background of hominid diversity also impinges on the matter of perspective that I have already raised. Clearly, *Homo sapiens* cannot be the perfected result of a single-minded process of fine-tuning over the eons, the peak of a single

NOT ALONE

BY IAN TATTERSALL

summit that we have laboriously climbed. Rather, our species is more like the sole surviving twig on a luxuriantly branching tree that represents a story of constant evolutionary experimentation—and, as often as not, of extinction.

The human story, the epic of *Homo sapiens* and of all its precursors and relatives who were not more closely related to today's great apes, begins in Africa some 6 to 7 million years ago. The origin of our family Hominidae is not very clear, partly because we really have no idea what the very earliest hominids *ought* to have looked like, and partly because the fossil record in the crucial 8-million- to 6-million-year period is very poor. The earliest current contender for hominid status is a form called *Sahelanthropus tchadensis*, known mostly from a cranium (skull minus the lower jaw) that was recovered in the central-western African country of Chad in 2001. This cranium caused much consternation when it was announced the next year, on the one hand, because in the case of certain features (for example the flatness of its face), it seemed to be a lot more advanced than one would have expected in a hominid thought to be between 6 million and 7 million years old. On the other hand, it did have an expectedly small braincase, below the size of a modern chimpanzee's.

The hominid status of *Sahelanthropus* is suggested by two features in particular. First, like us, it possesses fairly small canine ("eye") teeth which, unlike those of living apes, do not project in a fang-like manner. Second, its foramen magnum (the hole in the base of the skull through which the spinal cord descends to run through the vertebral column) is placed quite far forward. This suggests to many that *Sahelanthropus* was an upright walker, in whom the skull was balanced atop a vertically-held spine, rather than a quadruped in whom the spinal cord exited the back of the skull into a horizontal backbone. Here we have a critically important ingredient in recognizing *Sahelanthropus* as a hominid, because in recent decades, rightly or wrongly, upright bipedality has become pretty much universally accepted as the defining characteristic of our family.

Back in the time when *Sahelanthropus* lived, the African climate, once a humid one in which tropical forests flourished, had been slowly changing for several millions of years. A drying trend, and just as important a tendency toward marked seasonality of rainfall, had begun to affect the kind of vegetation the continent supported. Most significantly, the once-monolithic African rainforest began to break up and to become interspersed with patches of woodland and even grassland. Formerly diverse ape populations that had inhabited forests began to find themselves under stress as their habitat began to shrink, and some of them would have found themselves obliged to struggle along the forest edge rather than in deep forest, and even to move out into the more open woodlands that now lay adjacent.

The postures that the ancestral apes had adopted in the forest setting would have varied considerably. Some of them would have moved through the trees on the tops of branches, using mainly four-legged postures. Others would have preferred to suspend themselves from branches, and would habitually have held their bodies more vertically, especially while foraging. Some suspensory apes evidently found it easier to continue using vertical postures when habitat changes forced them at least part-time to the ground, either to forage there or to move from one clump of trees to another. These were the primates that ultimately became the ancestral hominids.

It is quite possible that several different kinds of primates made this particular kind of transition, and thus that not all of the forms that have been touted as human ancestors on the basis of presumed at least part-time bipedality were in fact hominids in the strictest genealogical sense. Certainly, it is possible that some experiments in bipedality as a response to environmental change ultimately failed. In any event, in the period between about seven million and four million years ago we have evidence of several different primates with claims to be an early biped and thus potentially to be the earliest hominid known. Besides *Sahelanthropus*, there is the perhaps marginally younger *Orrorin tugenensis*, from deposits that are almost exactly 6 million years old, in the Baringo Basin of northern Kenya. This form is known only from a few jaw and leg fragments. Although the most diagnostic parts of the skeleton are not preserved, it has been plausibly claimed on the basis of some portions of thigh bone that these fossils are indeed the remains of a biped. Also in contention are some bits and pieces of an Ethiopian form called *Ardipithecus*, with two claimed species, ranging in age from about five million seven hundred thousand years to four million four hundred thousand years ago. At the early end of this time range the argument for bipedality rests rather precariously on a single foot bone, while the younger *Ardipithecus* fossils include a fragmentary skull base that is said to show a relatively forwardly-positioned foramen magnum. Individually all these suggestions of bipedality are far from definitive, but put them all together and the pattern is quite suggestive. What's more, a stronger signal is one of diversity: if all or even most of these primates are indeed hominids in the strict genealogical sense, it means that from the very beginning of our family's history the human story has not been one of a steady trudge from primitiveness to perfection, but rather one of busy exploration of the many ways there evidently are to be a member of our family.

Better evidence for early bipedality comes from the Turkana Basin of Kenya, where a couple of sites about four million years old have yielded fossils of a form known as *Australopithecus anamensis*. Among these fossils are a piece of lower leg bone that includes the surface for the knee joint, and that demonstrates to many paleoanthropologists' satisfaction that hominids were up and walking on their hindlimbs at that time. But they were not walking upright exactly as we do. We know this best not from *A. anamensis* itself but from the closely related species *Australopithecus afarensis*, which is by far the most comprehensively documented of all of the early bipedal hominid species.

Australopithecus afarensis comes from sites in Ethiopia and Tanzania that date in the 3.8 million to 2.9 million year range. Fossils of this species include the famed "Lucy" skeleton (3.18 million years old) and several skulls and skull parts. They also probably include the renowned 3.5 million-year-old trackways from Laetoli, in Tanzania, which provide direct evidence that the hominids who made them were walking on two limbs. "Lucy" and other fossils of this kind demonstrate that these early bipeds were short-statured, standing only 1.1 m to 1.5 m tall. They also show that although they had broad pelvises (broader than ours; those of chimpanzees are very narrow), their limb proportions were unlike our own, with relatively short legs and long arms. Their shoulders were narrow and their hands and feet rather long:

ancestral features that would have stood them in good stead up in the trees. Indeed, while they clearly moved on two legs when on the ground and undoubtedly qualified as part-time bipeds, hominids of this kind were still agile climbers and probably regularly foraged and sought shelter in the trees. Because of this, and because, like today's great apes, they combined small brains with large, projecting faces, many paleoanthropologists have taken to calling these creatures "bipedal apes."

Although it is tempting to look upon a body form of this kind as a sort of "transitional" stage between ape-like tree-living and a human-like commitment to living on the ground, it was really nothing of the sort. Instead, it was a stable adaptation that served its possessors remarkably well for a very long time, even as a variety of species came and went that explored variations on this same basic theme. For both during and after the tenure of *Australopithecus afarensis*, several other hominid species came into being in various parts of Africa. From Chad once more, the 3.5 million-year-old species *Australopithecus bahrelghazali* (though possibly *afarensis*) has been described. In South Africa a recently discovered skeleton, maybe as much as 3.3 million years old, will almost certainly be ascribed to a previously unknown species of *Australopithecus*. In Kenya an entire new genus and species of a 3.5 million-year-old hominid, *Kenyanthropus platyops*, has recently been named. This form may have been at least broadly ancestral to the species *Homo rudolfensis*, known in the same region a million and a half years later. By 3 million years ago, the species *Australopithecus africanus* was well ensconced in South Africa, and the million years that followed witnessed not only a greater diversity of "gracile" (lightly-built) early bipeds than has yet been acknowledged with names (though newly published species include the possibly toolmaking 2.5 million-year-old Ethiopian species *Australopithecus garhi*), but also the appearance of the "robust" forms, equipped with huge grinding teeth and massive jaws. These first appear in East Africa about two and a half million years ago in the form of *Paranthropus aethiopicus*, which was followed by *Paranthropus boisei* in eastern Africa and *Paranthropus robustus* (and maybe also *Paranthropus crassidens*) in South Africa. Clearly, there was a major radiation in the 4 million to 2 million year period of early hominids of this general archaic "australopith" type. Once again, the clear signal in the hominid fossil record throughout this period is of diversity, the continent of Africa playing host to a variety of hominid species at every time period.

By something over a million years ago, however, the once-abundant australopiths disappear from the fossil record, almost certainly because they could not sustain competition with a new kind of hominid that began to show up about 2.5 million years ago. These hominids are usually classified as the first members of our own genus *Homo*, even though before 2 million years ago they probably looked more like australopiths. Behaviorally, however, the story was quite different. For while there was limited or no evidence that australopiths made stone tools, the making of such implements was the stock in trade of early *Homo*. This new innovation, the ability to produce a stone chip with a sharp cutting edge, must have revolutionized the lives of the early stone toolmakers, and this is a story that is amply fleshed out in the portraits of extinct *Homo* species that follow in this book. But following 2.5 million years ago another key innovation was yet to come. This was the acquisition of modern body proportions, something that probably occurred at just under 2 million years ago though it is best evidenced by the only 1.6 million-year-old "Nariokotome Boy" from northern Kenya, now usually classified in the species *Homo ergaster*. Here at last we have a hominid to which it seems legitimate to apply the adjective "human." The miraculously preserved Nariokotome Boy skeleton is that of a nine year old who stood about 159 cm tall when he died (early hominids developed more quickly than we do today), but who would have topped 185 cm had he lived to maturity. Here at last is a hominid who was recognizably modern in most of his bodily characteristics: a hominid who would have been thoroughly at home out on the expanding open savannas, far from the shelter of the trees.

This innovation evidently conferred on its possessors an altogether unprecedented mobility. For hard on the heels of the acquisition by *Homo ergaster* of modern body proportions (but before the development of more sophisticated stone tools or significantly enlarged brains), we find the first hominids outside Africa. The most spectacular of such finds is at the 1.8 million-year-old site of Dmanisi, in the Caucasus (Republic of Georgia), where the fossil hominids are relatively small-brained (though bigger than australopiths) and still possessed the simplest of stone tools. But there are also indications that *Homo* may have reached all the way to eastern Asia (China and Java) by that time, and certainly the famous species *Homo erectus* was already well established in the Javanese archipelago by well over one and a half million years ago. Europe, probably because of harsh Ice Age climates and formidable geographical obstacles, was colonized by humans only much later; the earliest archaeological indications in the subcontinent date from only about a million years ago, and the first hominid fossils, of the species *Homo antecessor*, are only about eight hundred thousand years old or thereabouts. While these earliest European hominid fossils may represent a couple of failed initial forays into the subcontinent, an indigenous group of hominids was well established in Europe by about half a million years ago, the most famous offshoot being *Homo neanderthalensis*, a talented and well-documented species that became extinct about thirty thousand years ago following the arrival on the European scene of *Homo sapiens*.

Where did *Homo sapiens* come from? At about six hundred thousand years ago in Africa, we find fossils of the species *Homo heidelbergensis*, which in some respects at least makes a plausible potential ancestor for our own species. This newcomer appears to have spread quite rapidly throughout the Old World; and it may, indeed, also have been the progenitor of *Homo neanderthalensis* in Europe. The hominid fossil record in Africa, following the appearance there of *H. heidelbergensis*, is not as good as we might wish; but by about one hundred and sixty thousand years ago something that cranially closely resembles our own species *Homo sapiens* is known from the Ethiopian site of Herto.

It is of interest that the earliest hominids, who *looked* just like us, do not appear to have *behaved* like us. We *Homo sapiens* today certainly have our anatomical peculiarities. But we are most sharply distinguished from other denizens of the living world by our behaviors, which are in turn underwritten by a unique cognitive system. Unlike any other organisms (as far as we know), we do not live in the world simply as nature presents it to us. Instead, we deconstruct and re-create the world around us via a mass of mental symbols. And it is in that mental world, of our own making, that we live. The Herto hominids are associated with a tool kit that was remarkably archaic even for the time; and indeed we do not find definitive archaeological evidence for symbolic behaviors until well under 100,000 years ago.

At Blombos Cave near the southern tip of South Africa, two very suggestive finds have been reported in archaeological strata dating between 80,000 and 70,000 years ago. These are of an ochre plaque bearing a definitely geometric design, which is probably the world's oldest known overtly symbolic object; and of pierced snail shells that were probably strung as a necklace or other bodily adornment (and body ornamentation, with its social and other connotations, seems to be a reliable indicator of symbolic mental processes). Some other similar expressions in Africa in the 50,000 to 60,000 year range also point to the nascent expression of symbolic processes in the continent, but it is not clear whether or not direct cultural transmission from such sources led to the most impressive of the known early manifestations of the human symbolic spirit.

These expressions come from Europe in the period following about 35,000 years ago, when the "Cro-Magnon" people, anatomically modern *Homo sapiens*, arrived in the area. Cro-Magnon culture was thoroughly drenched in symbol. These people decorated the walls of caves with some of the most powerful paintings ever made. They made delicate sculptures and engravings in mammoth ivory, bone, and antler. They made notations on small, flat bone plaques, some of them even maybe representing lunar calendars and terrestrial or astronomical maps. Cro-Magnons made music on simple bone flutes with remarkable sound capacities. Sometimes they buried their dead with an astonishing richness of grave goods and bodily decorations, suggesting some degree of social stratification as well as of division of labor. They almost compulsively decorated everyday functional items, such as scraper handles and spear throwers. There is no doubt that, as far as mental process was concerned, the Cro-Magnons were *us*.

What most plausibly happened is that, in the biological reorganization that resulted in the appearance of our species as a highly distinctive anatomical entity in Africa over 150,000 years ago, a new cognitive potential was born whose uses had to be released or discovered through the intervention of some cultural stimulus. A discovery of this kind would have been a pretty routine evolutionary phenomenon: after all, birds (and dinosaurs) had had feathers for warmth for millions of years before they co-opted them for flight. This process of recruiting old structures for new uses is often called "exaptation." The best candidate for a cultural stimulus of this kind is the invention of language. For, since it involves the manipulation of mental symbols to give them an infinite variety of meanings, language is a virtually complete metaphor for symbolic thought.

Whatever the case, the emergence of modern human intelligence does not appear to have resulted from a gradual process of fine-tuning of mental function over the ages, with a generation-by-generation improvement in mental performance as smarter individuals outbred the less intellectually endowed. It appears instead to have been a relatively sudden "emergent" event, in which a chance coincidence of a new biological acquisition with an existing mental substrate led to something with an entirely new potential. This potential was not fully explored immediately; instead, as you might expect, its expression was gradual and doubtless proceeded with many false starts and dead ends. Indeed, it can be argued convincingly that we are still in the process of finding out just what we can do with our exapted symbolic gifts (a notion that actually holds out some hope for the human future). But it is almost certainly this ability of ours for symbolic reasoning, together with our evident intolerance for competition (and we are still in the process of cleaning out our closest relatives, the great apes, from the rain forests of Africa and island Asia), that has led to our present (and entirely atypical) eminence as the only hominid on Earth.

This, then, is in brief the story of some 6 to 7 million eventful years of hominid evolution. But it is only the sketchiest of accounts. The fascination, as almost always, lies in the details, and the following pages reveal them. This book features a colorful cast of characters—or at least the twenty-two best-known of them—that have participated in the ongoing drama of human evolution over that period. In both word and image, it presents a series of intimate and unprecedented portraits of our extinct relatives. We actually get to *meet* an array of our precursors, presented as both actors and interactors in the great soap-opera of hominid history. Gary Sawyer and Viktor Deak and their gifted collaborators and consultants bring these long-vanished species alive with three-dimensional visual re-creations that are based on exacting science, but that are at the same time given reality, and even personality, by the exercise of extraordinarily skilful artistry. In turn, Esteban Sarmiento animates our extinct relatives with dramatic but scientifically informed and information-packed scientific stories, which are suffused with the reality that they were once living, breathing creatures. He follows all this with concise, up-to-date summaries of what is known of each species. This is, indeed, *the* guide-book to the human past, and the one that comes closest to being a personal time-machine. Read on!

TWENTY-TWO SPECIES OF

EXTINCT HUMAN ANCESTORS

BY ESTEBAN SARMIENTO

THE EARLIEST AFRICAN HOMINIDS

Sahelanthropus tchadensis • *Orrorin tugenensis* • *Ardipithecus ramidus* and *kadabba*

Due to limited information on the lifeways of these early hominids, there will be only one imaged narrative for all three fossil genera, and the one group claimed to have two species (*Ardipithecus*) will be treated as a single entry. Considering the scarcity of fossil evidence, this narrative will rely more on what is known of the early European and Asian ape fossils, especially *Oreopithecus*, which provide invaluable insights as to which great ape and human features are primitive and which are newly acquired. *Oreopithecus* suggests that features, such as the shape of the pelvis, lower back length, a knee joint with a carrying angle, a skull with a downward directed opening for the spinal cord, and non-interlocking canines, all of which have been used to classify fossils as hominids, are probably shared by the common human–African–ape ancestor, and do not distinguish a human lineage exclusive of African apes. As such, claims of human ancestry or human-like two-legged movement among the very early hominids based on these commonly shared features are speculative.

MORNING ENCOUNTERS

On reaching the crown of a yellow-wood tree the man-ape began bending back branches. He softly hooted to himself for there were no other man-apes in sight. Just when he felt the nest was right, he laid in it belly-up watching the sky darken, and waiting for the night. As the sun disappeared into the horizon, a small gust of wind kicked up from the east. The light drizzle that began shortly after the wind died prompted the man-ape to break back small branches with leaves and cover his body. Feeling comfortable with his new blanket, he quickly fell asleep.

In the middle of the night, the man-ape was awakened by the snap of a dried branch and the rustling of the forest's undergrowth. Alarmed, he left his nest, climbing up higher into the tree's crown, his eyes fixed in the direction from which the noise was coming. Some time passed without a stir. Relaxing, the man-ape dropped back into his nest and fell asleep.

The man-ape was awakened by the morning sun's rays peering above the horizon and warming the back of his head. From his vantage point he could see far out into the open valley below. He spied two hornbills flying into the crown of a fig tree, signaling the fruit was ripe. Climbing out of the nest, he quickly slid down the tree trunk as if it were a greased pole. He hit the forest floor with a thud and crashed through the undergrowth into the open valley.

The sight of a large cat staring at him from below a bush froze him in his tracks. He shrieked in panic, and clambered up a six-meter sapling at the forest edge. The cat gave chase, leaping into the tree and swatting at the man-ape with its forepaw. The tree bent under the weight of cat and man-ape, springing back when the cat was unable to secure a hold. Twice the cat jumped up into the tree, timing its leap with the tree's swaying. Twice it failed to bring down the man-ape, who now clung to the tip of the tree. Undeterred, the cat waited at the tree's base, staring at the man-ape above. Still screaming, the man-ape brought his knees up to his chest, simultaneously urinating and defecating. Sneezing and shaking its head and body free of the urine and feces, the cat ran off.

After cautiously looking around in all directions, the man-ape finally descended the sapling. Without his hands ever touching the ground, he walked along the forest edge in the direction of the fig tree, ever vigilant and reaching out to every large tree on the way. Coming as close as he dared to the forest edge, he dropped on all fours and sprinted across the open land until reaching the base of the tree. A troop of small monkeys was already in the crown of the fig tree feeding. Standing on two legs, the man-ape first fed from the figs on the lower branches. He then climbed up the tree and, having had his fill, sat on a crook staring up a ravine into a grove of fruit laden trees.

Resting until mid-morning, the man-ape awoke with new found enthusiasm. He began to call loudly in the direction of the tree grove. His calls reverberated throughout the valley and carried into the surrounding forest. They were soon answered by the calls of a man-ape group on the far side of the valley. A female man-ape's higher pitched calls grew closer and closer. Reaching the forest edge, the female, too, sprinted on all fours across the open land, climbed up the fig tree and sat next to the male.

They greeted, taking turns to stroke each other's faces with their index fingers. Feeling what appeared to be a small scab behind the female's ear, the male picked it off with his index finger and inspected it. Seeing it was a tick, the male brought it to his mouth, bit it with his front teeth, and swallowed it. The man-ape could now see the female's group in the distance. The male had followed the group for two weeks hoping to attract a female. By high noon the female's group grew closer. The two man-apes climbed down the fig tree and ambled up the ravine, stopping occasionally to eat the herbs that grew lush along a stream bank. In time the two would form a group of their own.

SAHELANTHROPUS TCHADENSIS

Skull, Teeth, and Diet

Sahelanthropus tchadensis is known from a single skull (TM 266-01-60-1), four lower jaw bone fragments (TM 266-02-154-1, TM 266-01-060-02, TM 292-02-01, and TM 247-01-02), and some individual teeth, composing a total of nine fossil specimens. Except for a missing lower jaw bone, the skull is complete, but it is cracked and deformed, and bone surfaces are abraded, obliterating detailed anatomy. The skull has an elongated shape with a relatively short, vertical face, which is set high relative to the skull vault. The eyes are set far apart from each other and have a very large continuous bony brow ridge above them. Brain size was first estimated at 320 cc to 380 cc, and later through computer imaging calculated to be 360 cc to 370 cc. This value is below the average for all three living great apes, but well within their range of variation. Most of the skull's teeth were broken off during fossilization. The last molar is the best preserved, with four other teeth providing only partial measurements. The lower jaw (TM 266-02-154-1), however, preserves most right lower cheek teeth and helps demonstrate that *Sahelanthropus* had a total of thirty-two teeth, as is common to all Old World monkeys, apes, and humans. Thick enamel ridges on the front teeth and canine, tooth length proportions, a short canine with a long root and wear restricted to its tip, a lower premolar which does not hone the upper canine edge, no gap between the lower canine and premolar, a short vertically set face, and a short skull base are all reminiscent of *Oreopithecus*, the best-known of the early great ape fossils. Similarities to *Oreopithecus* emphasize the fact that many of *Sahelanthropus'* human-like features are also shared by early great apes.

The foramen magnum (the skull opening for spinal cord passage) is positioned forward and oriented downward. *Sahelanthropus* has a nose opening that at midline reaches down to the bony palate as seen in gibbons and monkeys. By contrast humans and African apes have a nose opening that is set well above the mouth even at its midline. This suggests that the last common ancestor of African apes and humans would also have had the nose opening set well above the mouth, and that *Sahelanthropus* may predate this common ancestor.

As of yet there have been no tooth wear or carbon isotope analyses aimed at revealing *Sahelanthropus'* diet. Unfortunately most known teeth are heavily worn, providing very little clue. Its flooded grassland/woodland habitat suggests that it probably ate succulents, leaves, and the roots and bulbs of submerged grasses and herbs. Considering the Djurab Desert's long dry season, fruits would have been available only at certain times of the year. Large insects and small vertebrates when other foods were scarce probably supplemented the diet.

Skeleton, Gait, and Posture

Aside from the skull, jaw fragments, and teeth, there are no other known *Sahelanthropus* skeletal remains. Despite a human-like positioning of the foramen magnum suggesting to some that *Sahelanthropus*

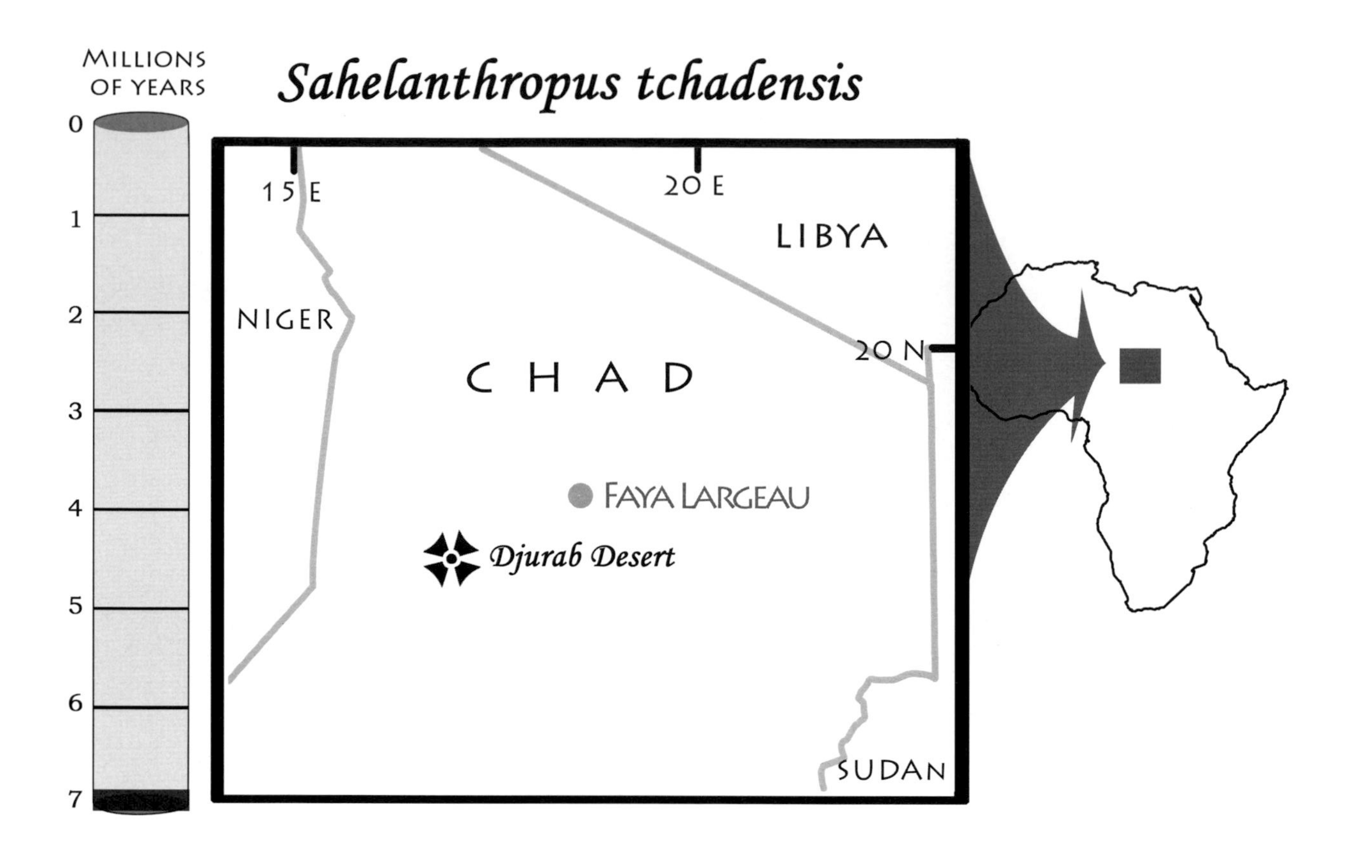

walked upright on two legs, without limb, hand, and foot remains it is not possible to verify its behaviors.

Fossil Sites and Possible Range

Sahelanthropus tchadensis fossils were found in localities 247, 266, and 292 of the Toros Menalla Formation (250 m above sea level) in the Djurab Desert of northern Chad. This formation documents cyclical contractions and expansion of a large inland lake, which in prehistoric times covered nearly the whole of Chad. This prehistoric lake was the precursor of the much smaller present-day Lake Chad. Deposited prior to this prehistoric lake's expansion, the lowermost part of the Toros Menalla is a four meter-thick layer formed from wind-blown sands. These sands represent prehistoric sand dunes, and record the early presence of a desert. This sand-dune layer is overlain by water worked sands containing clay and calcium carbonate from micro-organisms, all loosely cemented into a sandstone. This sandstone layer is approximately two meters thick and contains all the land living vertebrate fossils (*see* below, Animals and Habitats) found in the formation, including *Sahelanthropus*. It represents a lakeshore deposit of flood plains and small streams and lakeshore expansion over the underlying sand dunes. A well-layered deposit approximately a half meter thick, composed of fine clays and sands, containing fossil fish, covers this sandstone deposit. This clay-rich deposit represents lake-bottom sediments and records further lake expansion. Because lake expansion and lake retraction are cyclical, and at any one time and place (depending on distance from the lake center and lake size), sand dune, lakeshore, and lake bottom deposits are being laid down, different deposit types may be contemporaneous, while similar deposit types may not. It is difficult, therefore, to correlate deposits from different outcrops, unravel deposit sequence, and accurately date fossils.

Age

There are as of yet no absolute dates for *Sahelanthropus*. Comparisons of fossil animals found at locality 266 to those found in other African fossil bearing deposits of known age were used to arrive at a seven million year estimate. According to Michel Brunet and colleagues (the team that discovered the *Sahelanthropus* remains) fossil animals found at locality 266 are most similar to those found at Lothagam, a fossil bearing formation southwest of Lake Turkana, Kenya. The Lothagam deposit (i.e., Nawata Member; *see Ardipithecus,* Fossil Sites and Possible Range) with fossil animals that are most similar to those from Toros Menalla yields absolute radioisotope dates of 6.5 million to 7.4 million years. Considering that 1) Lothagam and Toros Menalla are separated by considerable distance (2800 km), 2) very few identified species are found in both deposits, and 3) Toros Menalla fossil animals compare most favorably to those from fossil sites in North Africa, the estimated dates for *Sahelanthropus* are at best very uncertain.

Tools

There are no tools found in the Toros Menalla deposits.

Differences Between Males and Females

There are no documented differences between males and females. The small canine and the absence of a gap between the canine and premolar suggest the skull is more likely female than male. Males may have had a honing lower premolar and a gap between the lower premolar and canine. Without additional remains, however, this is far from certain.

Animals and Habitats

Many of the animals found in the Toros Menalla deposits have yet to be identified. Of those species identified, nearly all belong to groups that are now extinct. These include giant true saber-tooth cats (*Machairodus*), an unidentified leaf-eating monkey, elephants with four tusks (*Anancus*, a gomphothere), a large water-loving relative of pigs (*Libycosaurus*, an anthracotheres), prehistoric giraffes (*Sivatherium*), three-toed horses (*Hipparion*), pigs (*Nyanzachoerus*), and an unidentified relative of musk ox. There are also three different hyena species, all abundant and representing genera that are now extinct. An aardvark, an African elephant, a ground squirrel, a porcupine, a pygmy hippopotamus relative (*Hexaprotodon*), various antelopes, and an otter all represent modern African genera. Of the two antelopes identified, the kob is a water-loving animal which usually inhabits swamps or flooded grasslands. The roan is distributed along watercourses, and prefers grasslands at the edge of forests or woodlands. Both antelopes are widely distributed in sub-Saharan African temperate zones.

The abundance of kob, roan, hippopotami, and anthracotheres (representing a majority of the mammal fossils found) and presence of gomphotheres, crocodiles, gavials, and pythons suggest large tracts of flooded grasslands and thin woodlands as occur along the Lake Chad shore today. Lungfish abundance suggests flood plains prone to seasonal droughts. Adjoining land not flooded by lakeshore or streams would have received little rain and supported a typical Sahel (desert shrub) vegetation. There are no exclusive forest-living mammals identified in the deposits. Most Toros Menalla mammals find their closest similarities to North African and European fossil counterparts, rather than to those from sub-Saharan Africa. These similarities suggest watercourses crossed the Sahara Desert near the Toros Menalla during the time *Sahelanthropus* lived.

Climate

Fayeu Largeau (235 m above sea level), a town about 260 km northeast of the Toros Menalla Formation, records monthly average temperatures that vary from a low 20.6° C (69° F) in December and January to a high 34.5° C (94° F) in May and June. May and June show a high mean monthly maximum temperature of 44.5° C (112° F). December and January show a low mean minimum monthly temperature of 13.3° C (56° F). Reflecting a desert climate marked by a sizeable difference between day and night temperatures, mean maximum and minimum monthly temperatures may differ by as much as 18° C (32° F). Average annual rainfall is 15 mm, with all rainfall falling June through September, and two-thirds of the annual total falling in August. Towns at similar latitudes as the Toros Menalla deposits—Shendi, Sudan, and Agadez, Niger—also record relatively low rainfalls (100 mm and 300 mm) with the same annual pattern. With an African continent that has changed little in size and relative position to the Equator since the time *Sahelanthropus* lived, there is no reason to expect that rainfall and rainfall patterns would have been much different from what they are today. The presence of sand dunes in the Toros Menalla deposits prior to lake expansion attests to this fact. During *Sahelanthropus* times, water availability in this area was more than likely the result of changes in drainage patterns as opposed to changes in rainfall pattern.

Classification

Given its known similarities to the early great ape *Oreopithecus, Sahelanthropus* was probably a member of the human–great ape group. Because the wrist and forearm anatomy presents the most reliable evidence as to whether a fossil is a member of the human–great ape group, without this evidence inclusion of *Sahelanthropus* into this group is not definite. Alternatively, its non-honing canine premolar complex and skull features, suggest *Sahelanthropus* may be a member of the human lineage, post African–ape divergence. Lacking definitive evidence, this conclusion is also speculative. Considering its geologic age difference and geographic distance from the less complete *Orrorin* and *Ardipithecus*, a separate species and generic designation for *Sahelanthropus tchadensis* is well justified.

Historical Notes

All described *Sahelanthropus* remains were found between July 2001 and February 2002. Djimdoumalbaye Ahounta is credited with finding the *Sahelanthropus* skull in July of 2001. A year later, in a July 2002 *Nature* issue, Michel Brunet and a long list of colleagues (including Yves Coppens, who also described *Orrorin*, and David Pilbeam, who proposed *Ramapithecus* as an exclusive human ancestor in the nineteen sixties and seventies) designated the skull as the holotype specimen (i.e., the fossil that the describer chooses to represent the species), creating a new genus and species *Sahelanthropus tchadensis*. In an October 2002 *Nature* issue, Milford Wolpoff, Bridget Senut, Martin Pickford, and John Hawks argued that *Sahelanthropus* was not a habitual biped and therefore not an exclusive human ancestor. In an April 2005 *Nature* issue, Brunet and colleagues reported three additional *Sahelanthropus* fossils and new estimates for this fossil's brain size. The new fossils showed the absence of a canine premolar honing complex. *Sahelanthropus* literally translates as "man from the Sahel." The latter is the name of the African region bordering the southern Sahara Desert. The species name "*tchadensis*" was assigned in recognition of all fossil specimens recovered in Chad. The nickname "Toumai" given to the *Sahelanthropus* skull means "hope of life" in the local Chad language (Goran), and is a name commonly given to human babies born just before the dry season. All *Sahelanthropus* fossils are housed in the Departement de Conservation des Collections, Centre National d'Appui à la Recherche (CNAR) in N'Djamena, Chad.

The enigmatic *Sahelanthropus tchadensis* contemplatively surveys the African landscape some seven million years ago.

ORRORIN TUGENENSIS

Skull, Teeth, and Diet

Two jaw fragments with three molars from a single individual (BAR 1000'00) and some non-associated teeth (i.e., the lower molar found in 1974, a lower premolar, an upper canine, an upper front tooth, and right and left upper wisdom teeth) discovered in Kenya are all that is known of the *Orrorin tugenensis* skull and teeth. The molars have low rounded cusps and thick enamel, and the upper front tooth is large and spatula-shaped. The upper canine is short and pointed with a shape reminiscent of that of living female chimpanzees. Given its length, it could not have protruded very far past the tooth row. A front, longitudinal groove on the canine is suggestive of a possible honing function, but can also be interpreted as an ancestral remnant, i.e., a feature that no longer serves its original function, but is inherited from an ancestor, in which this function was served.

Unfortunately, the sex of this early hominid is unknown. If the fossils belong to a female, the male canine could have been much larger and may have had a honing function, as may occur in male and female living great apes. Overall the teeth are claimed to have thick enamel and to be small relative to body size, but enamel thickness has not yet been fully measured. Because remains come from different time intervals and most teeth are not associated, relative tooth size estimates are uncertain, i.e., these estimates may be reflecting differences in body sizes over time and not necessarily relative tooth proportions. With a pointed canine, low rounded molar cusps, and large upper front teeth, *Orrorin* probably ate fruits and could fall back on seeds or other hard items when fruits were not available. The fact that *Orrorin* inhabited an equatorial area with relatively heavy rainfall where fruits would have been available year round supports such an interpretation.

Skeleton, Gait, and Posture

A finger bone (BAR 349'00), a right arm bone shaft (BAR 1004'00), a small thigh bone fragment (BAR 1215'00), the upper two-thirds of two left thigh bones, one with (BAR 1002'00) and one without joint surfaces (BAR 1003'00), and more recently an end finger bone (BAR 1901'01) represent the totality of *Orrorin tugenensis* skeletal fossils so far unearthed. Martin Pickford and the discoverers of these fossils point to the thicker cortical bone on the lower half of the thigh bone neck when compared to the upper half as evidence that this species walked on two legs. Baboons and other four-legged ground-moving primates, however, have a similar cortical bone distribution, suggesting that there is no exclusive association between this distribution and two-legged ground movement.

The shape of the arm-bone shaft indicates that the upper limb was probably used in weight bearing. The finger bone is similar in its dimensions and curvature both to that of living baboons and to that of *Australopithecus afarensis* (A.L. 288-1, Lucy), which suggests an animal that moved on the ground, but also used trees. Unfortunately, the finger bone is an isolated find, which may be three to four hundred thousand years younger than the

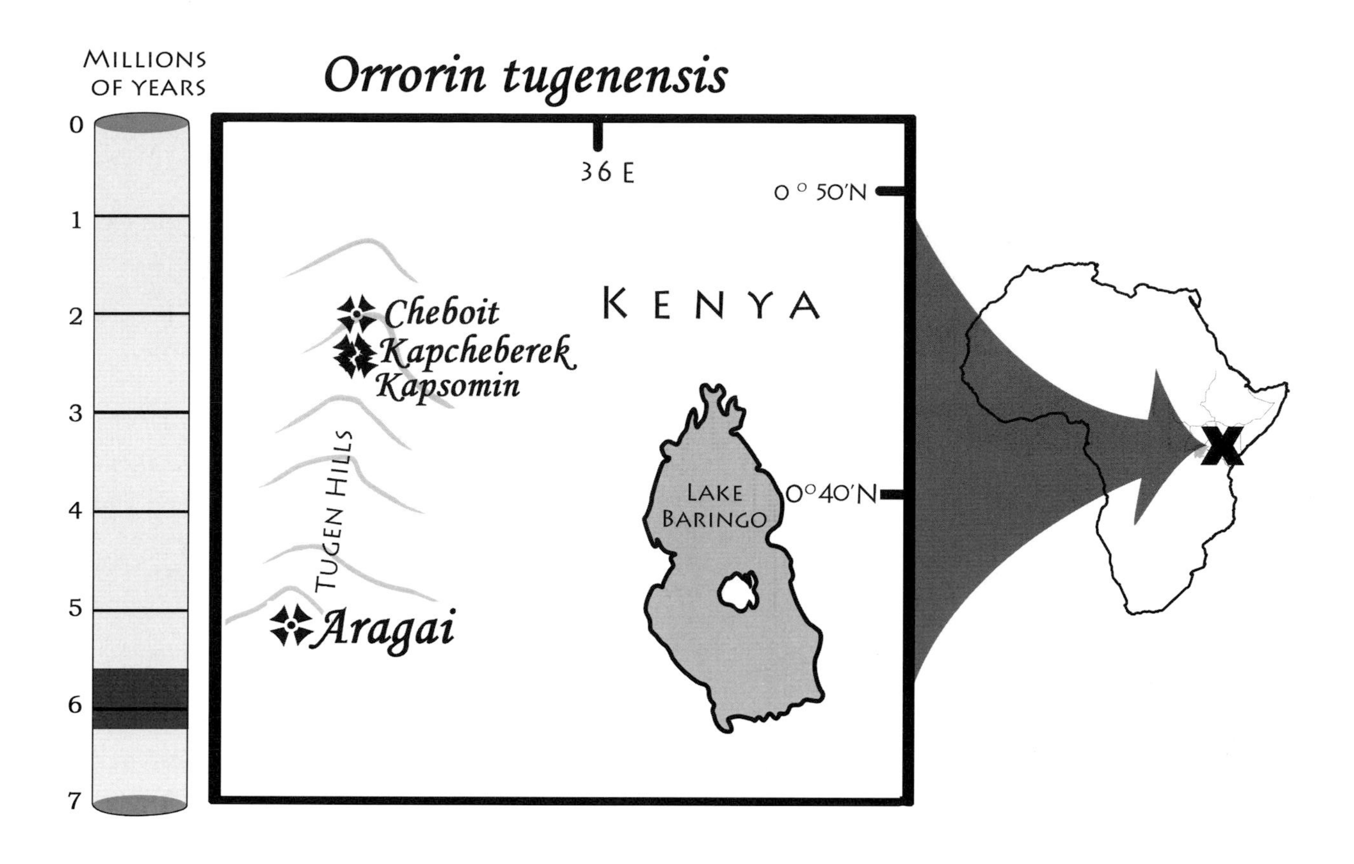

other *Orrorin* fossils, and it is not clear that it actually belongs to *Orrorin*. The assumption that over this time interval *Orrorin* is the only early hominid living in this area appears to be the only reason for referring the finger bone to this species. With large leaf-eating monkeys known from other contemporaneous deposits, the possibility that the finger bone is from a monkey cannot be discounted. Except that it comes from Kapsomin (where most *Orrorin* remains, including the teeth, were found), much the same applies to the end finger bone. In fact, there is no good evidence that this bone represents the end thumb bone as opposed to one from the lateral fingers, or even that it was not from a toe. All in all, there is currently precious little evidence bearing on how *Orrorin* moved.

Fossil Sites and Possible Range

Orrorin tugenensis has been reported from the Lukeino Formation at four different localities: Cheboit, Kapsomin, Kapcheberek, and Aragai (from north to south) along the eastern foothills of the Tugen Hills west of Lake Baringo, Kenya. The Lukeino Formation is a Late Miocene (5.3 million to 7 million years ago) shallow lake and stream deposit composed mainly of silts, sands, and thin limestone layers. The formation rests on a volcanic flow and is capped by basalt. A short time after its deposition the Lukeino Formation was intruded on by a basalt flow (i.e., dolerite sill), which separated it into upper and lower parts. Depending on locality, the *Orrorin* fossils come from different levels within the formation, reflecting different geologic ages. In order of decreasing depth and age, *Orrorin* fossils from Aragai, Cheboit, and Kapsomin are all found below the basalt intrusion. Fossils from Kapcheberek are found above the basalt intrusion in the highest levels of the formation. Most *Orrorin* fossils come from the Kapsomin outcrops, which represent a shallow lake deposit of very slow moving water. Because they have not been rolled along a stream bed or abraded by water action, these fossils preserve fairly good detail.

Age

The volcanic flow on which the Lukeino Formation rests, and the basalt capping it, provide radioisotope dates of 6.2 million and 5.65 million years. These dates represent a maximum and minimum age for when the formation's deposition began and ended. Radioisotope dating of the basalt intrusion is slightly younger (30,000 years) than the capping basalt. Considering that the *Orrorin* fossils from Aragai are found in the lowest levels of the deposit and those from Kapcheberek in the highest levels, their age differences could be as much as half a million years.

Animals and Habitats

Many of the animals found in the same deposits as *Orrorin* have not yet been fully identified, and the deposits as a whole have not yet been fully sampled. Considering an equatorial habitat, greater diversity of animals than those reported would be expected. The identified animals are associated with habitats with much more plant growth than those in which later hominids (i.e., *Australopithecus*) were known to live. For instance, the presence of forest duikers and pygmy or dwarf antelopes suggests forests and heavy undergrowth. There are no members of the hartebeest/wildebeest family, antelopes usually found in open country. Instead there are sitatungas, kobs, and impalas. Impalas often inhabit woodlands and the other two wet grasslands and marshes. Wet grasslands are also indicated by the presence of white rhinoceroses and cane rats. Mole-rats and blesmoles are typical to mosaic habitats with woodlands, forests, and clearings with grass or dense growth. Reflecting the incomplete sampling of the deposit, there is only one small leaf-eating monkey known. This number is far short of the six monkey species found in the area today. Overall the animals indicate a mosaic of forests and woodland habitats with dense undergrowth, and wet grasslands along the lake margin. Other deposits, close in age and nearby to those in which *Orrorin* is found, preserve tree trunks that further indicate forest patches.

Climate

Eldoret, a town on the western edge of the eastern Great Rift Valley (2132 m above sea level), and approximately forty kilometers west of the Aragai locality, receives on average 1232 mm of rain per year. Rainfall is seasonal, with most rain (880 mm) falling in March through September. January and December are the driest months, receiving a combined average rainfall of 80 mm. Monthly average temperatures fluctuate between 16.1° C and 18.9° C (61° F and 66° F), with the coldest averages occurring in July and August and the warmest in February and March. On rare occasions October temperatures may reach freezing. With the exception of Aragai, all *Orrorin* localities are about one thousand meters' elevation lower than Eldoret, and would have temperatures that are approximately 6.5° C (12° F) higher. Four hundred meters higher than the other localities, Aragai would show temperatures only 4°C (7° F) warmer than Eldoret.

Today annual rainfall around the *Orrorin* fossil localities is just below the amount associated with tropical rainforests and supports a mosaic of woodlands and bushlands. It is possible that the deposits may have been at elevations as low as Lake Baringo (970 m) and would have received somewhat less rain. Ongoing mountain-building may have been associated with local differences in rainfall during the time *Orrorin* lived when compared to today. Otherwise, it is difficult to imagine why the overall climate would have been much different during the time *Orrorin* lived than it is today.

Morning sunlight bleeds through the canopy of a humid jungle to reveal *Orrorin tugenensis* rising to his feet within the secret confines of a primordial forest.

Classification

Few and incomplete remains make classification of *Orrorin tugenensis* uncertain. Placing it in a new species and genus, however, appears to be a reasonable measure. As previously noted, because *Orrorin* is as much as a million years removed in time and/or separated by considerable distances from other possible early hominid fossils (i.e., *Sahelanthropus* and *Ardipithecus*), it is highly unlikely that it belongs to the same species or even the same genus as either of these. Nevertheless, given its fragmentary remains, its describers had difficulties pointing out convincing features that would merit generic separation from the Ethiopian *Ardipithecus*. While such defining features probably do exist, they must await more complete remains and analyses.

The fragmentary nature of *Orrorin* remains also precludes its uncontested inclusion in the human lineage. On one hand, some early fossil apes that are clearly not human ancestors have relatively short canines that do not protrude past the tooth chewing surface. In addition, because female apes may also have short non-interlocking canines, they, too, can be confused with human ancestors. On the other hand, if further evidence is unearthed demonstrating that *Orrorin* walked on two legs, this hominid would be one of the earliest, if not the earliest, member of the human lineage to do so.

Historical Notes

In 1974, Martin Pickford found the first molar of *Orrorin tugenensis* at Cheboit, Kenya, reporting it in *Nature* the following year. Twenty-six years later in October and November of 2000, a French-Kenyan team returned to the Tugen Hills and found all of the other *Orrorin* fossils. In a 2001 *South African Journal of Science* issue, Pickford and Bridget Senut announced the new *Orrorin* finds and asserted that this early hominid would have routinely moved around on two legs. At the end of January of that same year in *Comptes Rendus de l'Académie des Sciences,* Senut, Pickford, Dominick Gommerey, Pierre Mein, Kiptalam Cheboi, and Yves Coppens, based on these same finds, erected a new genus and species *Orrorin tugenensis*. Cheboi is credited with uncovering the fragmentary mandible that was described as the holotype specimen. "*Orrorin*" means "original man" in the Tugen language. The species name "*tugenensis*" refers to the Tugen Hills where it was found. All *Orrorin* fossils are housed in the National Museums of Kenya, Nairobi, Kenya.

ARDIPITHECUS RAMIDUS AND KADABBA

Skull, Teeth, and Diet

An upper and lower set of associated teeth, a lower jaw bone with a milk molar and front tooth, an upper jaw bone with a molar and canine, nine isolated teeth or teeth fragments, a very incomplete skull base, and a skull wall fragment are all that is known of the *Ardipithecus ramidus* skull. As in living great apes, the *Ar. ramidus* first milk molar (ARA-VP-1/129) has a single large dominant cusp, lacks a sizeable crushing surface, and is narrow from side to side. In its dimensions the milk molar most closely matches that of living chimpanzees. In adult *Ar. ramidus* the front teeth are narrower relative to the cheek teeth than those of chimpanzees, and in this ratio more closely approximate gorillas. The canines protrude past the cheek tooth chewing plane and are larger relative to the cheek teeth than in all *Australopithecus* species. They are close in relative size to those of female great apes. As is sometimes seen in female great apes, there is no gap in the tooth row between the lower canine and premolar, both upper and lower canines show tip wear, and neither canines nor lower premolars show wear associated with canine sharpening.

The *Ardipithecus ramidus* first premolar has a single large dominant cusp resembling those of chimpanzees. The molars are much smaller than those of *Australopithecus afarensis*, and molar enamel has a thickness intermediate in value between the thin enamel of chimpanzees and the thick enamel of *A. afarensis*.

The *Ardipithecus ramidus* skull base fragments (ARA-VP-1500) show a flat jaw joint surface, as seen in chimpanzees. The skull base is too incomplete to determine with certainty whether its foramen magnum showed a human-like or an African ape–like orientation and position (*see Sahelanthropus*). The small joint surfaces corresponding to the first neck bone indicate a relatively small head.

The diet of *Ardipithecus ramidus* has yet to be studied. Molar enamel thickness intermediate between that of *Australopithecus afarensis* and chimpanzees, and molars smaller than those of A. *afarensis*, indicate that it probably ate less abrasive foods with less variety than those eaten by *A. afarensis*. Although upper front tooth width greater than in *A. afarensis* but less than in chimpanzees suggests it may have eaten less fruit than chimpanzees, it would have eaten more ripe fruit, succulent plant parts, and immature leaves than *A. afarensis*.

All that is known of the *Ardipithecus kadabba* skull and teeth is a lower jaw fragment with associated front tooth, canine, premolar, and three molars from Alayla (ALA-VP-2/10), two isolated lower molars and a lower canine from Saitune Dora, and seven isolated teeth (two lower premolars, two upper molars, a right upper canine, and two upper premolar fragments) from Asa Koma. Molar size and tooth enamel thickness are within the range of *Ar. ramidus*. Canine size and shape (long and pointed) suggest the tip of this tooth protruded past the cheek-tooth chewing surface. The upper canine (ASK-VP-3/400) does not show tip wear, but shows a honing facet, in front along its length, corresponding to the lower canine. The lower premolar (ASK-VP-3/403) has a more asymmetric outline (wider from front to back on the cheek side and narrower on the tongue side

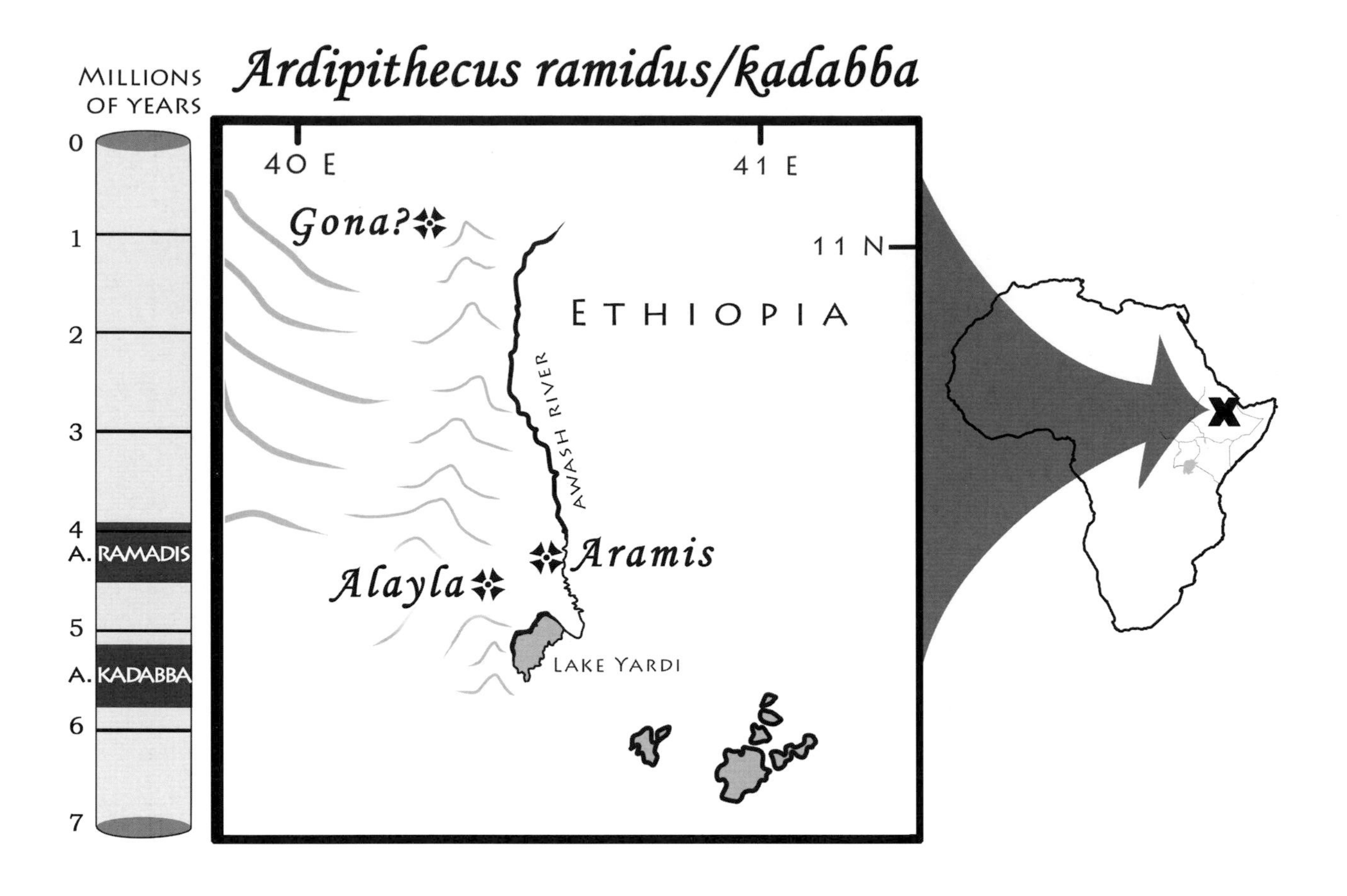

of the tooth) than that of *Ar. ramidus*, and in this aspect is similar to great apes. The lower canine also has a long wear facet where it meets the upper canine. As in great apes, therefore, *Ar. kadabba* had self-sharpening interlocking canines. Similarities to *Ar. ramidus* in what few teeth are known suggests that *Ar. kadabba* had a similar diet.

Skeleton, Gait, and Posture

The *Ardipithecus ramidus* arm bone shaft, and the associated but incomplete arm and forearm bones (ARA-VP-7/2), have yet to be described. Although based on the published photograph, the arm and forearm bones seem to show a number of features that suggest a four-legged ground moving ape as opposed to one that climbed trees (i.e., an elliptical shoulder joint, arms longer than forearms, and a wide wrist joint), this is by no means certain. The arm bone joint surface suggests *Ar. ramidus* was about thirty percent larger in size than Lucy (A.L. 288-1).

It is not certain if the upper limb, hand, and foot bones assigned to *Ar. kadabba* belong to this species. With an age possibly half a million years younger than the other *Ar. kadabba* remains, the complete toe bone from Amba East (AME-VP-1/71) may very well belong to another species. Its length to shaft-width ratio and joint orientations strongly suggest a grasping foot. Except for this toe bone, the only other fossils to preserve joint surfaces are the lower half of two supposed partial finger bones (ALA-VP-211 and DID-VP-180) from Alayla and Digiba Dora. The Alayla bone may, in fact, be a toe bone. The Digiba Dora finger has a lower joint orientation that suggests the fingers were strongly flexed. Toe and finger bone fragments, however, are notoriously difficult to identify, and given that they are isolated finds may have belonged to apes or monkeys living in the area. The rest of the remains assigned to *Ar. kadabba* (i.e., one-third of a collar bone shaft [STD-VP-2/893], two arm bone shafts, one isolated [ASK-VP-3/78], and one with an associated forearm bone shaft [ALA-VP-2/1010]) lack joint surfaces and have limited use in reconstructing *Ar. kadabba* behaviors. Arm bone shaft diameter suggests a body size approximately like that of Lucy (A.L. 288-1). As is, there is very little evidence bearing on how *Ar. kadabba* moved about.

Fossil Sites and Possible Range

Ardipithecus ramidus fossils are found in stream and lake silt deposits (at 625 m above sea level) in the Aramis Member of the Sagantole Formation on the west side of the Middle Awash River Valley, between the upper Adgantole River and lower Sagantole River drainages in Ethiopia. These deposits are exposed by the Aramis River headwaters from where the fossil locality gets its name. Aramis is situated in the Afar rift about 246 km northeast of Addis Ababa, Ethiopia. Except for a left arm bone and corresponding forearm bones (ARA-VP-7/2), all *Ar. ramidus* remains are sandwiched between volcanic ash sediments (i.e., the Daam Aatu basaltic tuff above and the Gaala vitric tuff below), which enables radioisotope dating.

Ardipithecus ramidus is also believed to be found in the As Duna deposits at Gona 32 km west of Hadar, Ethiopia (*see Australopithecus afarensis*). Unfortunately, the Gona fossils, as well as those Kenyan fossils from the Apak Member of the Nachukui Formation in Lothagam, southwest of Lake Turkana, and in the Tugen Hills (Chemeron and Tabarin; *see Orrorin tugenensis* and *Paranthropus aethiopicus*) are too incomplete to classify. Comparable geologic ages are the main reason for suspecting these could belong to *Ar. ramidus*.

Ardipithecus kadabba fossils are found in silty clays in the Kuseralee and Asa Koma Members of the lower Sangatole and Adu-Asa Formations. They are also found in silty clays at Digiba Dora, Saitune Dora, Alayla, and Amba East localities. All these localities are found on the western margin of the Afar rift between 690 m and 800 m above sea level, just northwest of Lake Yardi, Ethiopia with the exception of Amba East (650 m above sea level), which is found directly north of the lake. The Alayla site is 217 km northeast of Addis Ababa, 30 km southwest of the *Ardipithecus ramidus* site, 24 km southeast of the Amba East site, 4 km west of Saitune Dora, and 14.5 km south-southeast of Digiba Dora. The Asa Koma and Adu Dora sites are found more or less between Alayla and Digiba Dora. All *Ar. kadabba* sediments represent lakeshore and small stream deposits. Interlayered volcanic sediments enable both radioisotope dating and correlation of deposits to each other. Unfortunately, correlation and dating of the relevant layers has yet to be reported on.

More complete remains than the two isolated molars with a possible date of 5.32 million years known from the upper Nawata Member of the Nachukui Formation at Lothagam, southwest of Lake Turkana, Kenya, may show that *Ardipithecus kadabba* also comes from these deposits.

Age

The Gaala vitric volcanic ash (tuff) that underlies all the *Ardipithecus ramidus* fossils was originally dated at 4.39 million years but subsequently revised to 4.42 million years. The Daam Aatu layer of basaltic volcanic ash overlying the remains appears to yield the same date, indicating that the four meters of fossil bearing sediments sandwiched in between the two layers of volcanic ash was deposited over a relatively short time. All the *Ar. ramidus* fossils with the exception of the forearm bones (ARA-VP-7/2) are, therefore, 4.42 million years old. The forearm bones are younger than 4.42 million years, but probably much closer to this age than to an overlying volcanic ash (tuff) dated at 3.9 million years. At Gona, a layer of volcanic basalt between fossil deposits yielded an absolute date of 4.15 million years. Fossils found below the basalt may be as much as a million years older than those found above the basalt, creating further uncertainty as to their *Ar. ramidus* classification.

Layers of volcanic ash, sandwiching the *Ardipithecus kadabba* fossils in the lower Sangatole Member, were dated at 5.55 million and 5.18 million years, suggesting *Ar. kadabba* fossils from Amba

East have an age somewhere within these two dates. Volcanic ash in the Asa Koma Member at Asa Koma and at Digiba Dora, and basalt at Saitune Dora, all overlying *Ar. kadabba* fossils, was dated at 5.63 million to 5.57 million, 5.68 million, and 5.54 million years. At Alayla *Ar. kadabba* fossils overlie basalt dated at 5.77 million years, which supposedly corresponds to the base of the Asa Koma Member. *Ar. kadabba* fossils from the Asa Koma Member are, therefore, between 5.77 million and 5.57 million years old, and possibly as much as 590,000 years older than those from the lower Sagantole Formation at Amba East. The *Ar. kadabba* fossils at Digiba Dora and Saitune Dora have a minimum age comparable to the Asa Koma fossils, but their maximum age is as of yet undetermined, and they may be much older. The Alayla fossil is clearly younger than 5.77 million years, but a minimum date is still uncertain.

Tools

There are no tools found in either the *Ardipithecus ramidus* or *Ardipithecus kadabba* deposits.

Animals and Habitats

Of those identified fossil mammals found with *Ardipithecus ramidus* many belong to extinct genera, i.e., the dirk-tooth cat (*Megantereon*), prehistoric elephants with lower jaw tusks (*Deinotheirum*), four-tusked elephants (*Anancus*, a gomphothere), an extinct cousin of modern day bears (*Agriotherium*), two extinct pig species (*Nyanzachoerus*), an extinct otter (*Torolutra*), a three-toed horse (Hipparion), a prehistoric baboon (*Parapapio*), and a leaf-eating monkey (*Kuseracolobus*). Others belong to genera that today are no longer found in Africa, i.e., raccoon dogs (*Nycteruetes*), Indian bush rats (*Golunda*), and two species of soft-furred rats

An agitated troop of *Ardipithecus ramidus* charge into a shallow stream to ward off an approaching predator some four and a half million years ago.

(*Millardia*), reflecting an Asian connection. The remaining mammals—hyena, dwarf mongoose, white rhinoceros, pygmy hippopotamus, giraffes, dwarf or pygmy antelopes, kudus/bushbucks, and African mole-rats—are all modern genera still found in Africa today. Leaf-eating monkey abundance (30 percent of all fossil animals found) and the presence of kudus/bushbucks were used to argue that the deposits documented closed woodlands conditions. Overall the fossil animals suggest a habitat mosaic of woodlands with thick underbrush, flooded grasslands, and swamps. Presence of African mole-rats (*Tachyoryctes*), an animal that today is found as low as 500 m above sea level, does not show deposits formed at elevations higher than the 625 m above sea level where they are found today. Dwarf mongooses, however, suggest that the deposit probably did not form at elevations higher than 1600 m above sea level.

Very few animals found with *Ardipithecus kadabba* have been reported on. Most animals are also found in the *Ardipithecus ramidus* deposits, suggesting the habitats of these two species were more or less similar. Although absence of hartebeests and gnus was interpreted to indicate there was no open grassland, this absence may also reflect the early age of the deposits, prior to when this animal group spread and diversified. The presence of cane rats (*Thrynomys*) indicates flooded grasslands with at least an annual rainfall of 500 mm. There are no exclusive forest animals found in the deposits of either *Ar. ramidus* or *Ar. kadabba*.

Climate

The climate in the Afar depression is highly variable. Generally, rainfall is seasonal, with the months of December through February receiving the least rainfall and June through September the most. Rain increases with increasing elevation. Over a period of thirty-seven years, the Kembolcha weather station (1920 m above sea level and 100 km northwest of Aramis) reported 1056 mm of rain annually. More than half of the rain (655 mm) fell July-September. November-January received only 61 mm of rain. The temperature varies relatively little year round, with the highest mean monthly temperature recorded in June 22.8° C (73° F) and the lowest in December 15.6° C (60° F). Towns at lower elevations receive less rainfall but have warmer temperatures. The *Ardipithecus kadabba* and *Ardipithecus ramidus* localities would see a total 40 mm of rainfall in November-January and about 350 mm in July-September. Mean annual rainfall would be close to 600 mm but can fluctuate from year to year between 770 mm and 440 mm. At nearly one thousand meter lower elevations than Kembolcha, the highest mean temperatures are about 30° C and the lowest temperature 22° C. At ten degrees latitude north of the Equator, it is unlikely that the rainfall pattern was less seasonal in *Ardipithecus* times. Although higher elevations would have been associated with more rainfall, there is little evidence that this area was at much higher elevations than it is at present.

Classification

With as much as 1.1 million years separating *Ardipithecus ramidus* from most *Ardipithecus kadabba* fossils, the two probably are different species. Their fossil remains, however, are so incomplete that none of the reported features distinguish them as different species. Differences in canine premolar honing do not convincingly do so. Great ape females (most often orangutans, but also gorillas) do not always show interlocking and self-sharpening canines. In fact, this absence had many paleoanthropologists in the nineteen sixties and seventies believing that females of an early Asian ape species allied to living orangutans were human ancestors, and were members of a different genus (*Ramapithecus*) than the males (*Sivapithecus*) of the same species. Differences in development of interlocking canines in *Ar. kadabba* vs. *Ar. ramidus*, therefore, could just as likely represent male and female differences.

No anatomical features in *Ardipithecus ramidus* or *Ardipithecus kadabba* definitively demonstrate that either or both are an African ape, a member of the common human–African ape lineage, or a hominid. Even if one makes the likely assumption that non-interlocking and non-honing canines is the primitive condition for the common human–African ape ancestor, and thus *Ar. kadabba* exclusively shares the more recently evolved interlocking self-sharpening canines with chimpanzees and gorillas, canine development is just too variable a feature on which to base evolutionary relationships. The same applies to other *Ar. kadabba* and *Ar. ramidus* chimpanzee-like features, i.e., thin enamel, milk molar shape, and a chimpanzee-like toe bone (if one assumes it actually belongs to *Ar. kadabba*). During evolution of the human–African ape lineage, these features may have fluctuated many times between the chimpanzee-like and human-like condition, so that they need not reflect a shared heritage with any one lineage. Arguments aimed at showing *Ardipithecus* is a human ancestor, by linking this fossil through a perceived reduction in canine size over time to *Australopithecus*, are also problematic. The presence of Asian mammals in the *Ardipithecus* deposits (e.g., a racoon dog and a number of rats) raises the possibility that *Ardipithecus* could represent a lineage of Asian apes that reached Africa and later became extinct.

Finally, it is not clear why *Ardipithecus* should have a different generic name than *Australopithecus* if, as claimed, the *Ardipithecus* arm bones (ARA-VP-7/2) and the *Australopithecus anamensis* teeth (ARA-VP-14/1) represent a direct lineage separated by as little as 220,000 years. In fact, there are no features distinguishing the very incomplete remains of *Ardipithecus* and those of *A. anamensis* at the generic level, and the two would best be assigned to the genus *Ardipithecus*. It remains to be seen whether enamel thickness, the only feature distinguishing *Ar. ramidus* and *Ar. kadabba* from *A. anamensis*, is the result of population variation. If this turns out to be the case, their species differences would also be brought into question.

Historical Notes

Gen Suwa is credited with finding the first remains of *Ardipithecus ramidus*, an upper molar (ARA-VP-1/1), in December 1992. In December 1993, Gada Hamed found the *Ar. ramidus* holotype specimen (ARA-VP-6/1), comprising ten associated teeth. Based on this holotype specimen, Tim White, Suwa, and Berhane Asfaw in a September 1994 issue of *Nature* described a new species, *Australopithecus ramidus*. In a May 1995 corrigendum (correction) to *Nature*, the same authors, without presenting any additional evidence, changed this species genus to *Ardipithecus*. They also announced a newly found mandible with a partial postcranial skeleton; these remains have yet to be described. In 1997, Y. Haile-Selassie found what would become the *Ardipithecus kadabba* holotype specimen, a right lower jaw fragment (ALA-VP-2/10), preserving the last molar and later nearby four lower isolated teeth believed to belong to the same individual. In a July 2001 *Nature* issue, Y. Haile-Selassie described the subspecies *Ardipithecus ramidus kadabba*, making the ALA-VP-2/10 jaw fragment the holotype specimen. In March 2004, Haile Selassie, Suwa, and White elevated *Ardipithecus ramidus kadabba* (subspecies) to *Ardipithecus kadabba* (species) based on six newly found isolated teeth that evidenced interlocking and self-sharpening canines. In an April 2006 *Nature*, Tim White and twenty-one co-authors argued, on the basis of canine size and function, that a newly reported fossil believed to be *A. anamensis* evolutionarily linked *Ardipithecus* to *Australopithecus afarensis*, reputedly establishing *Ardipithecus*' membership in the human lineage.

"Ardi," the prefix for this fossil's generic name, means "ground floor" in the Afar language. The species name "ramid" means "root" in the Afar language. It was chosen as a species name in recognition of the Afar people, and their strong roots in the Middle Awash River area. The species name "*kadabba*" means "basal ancestor" in the Afar language. All *Ardipithecus* fossils are housed at the National Museum of Ethiopia, Addis Ababa.

OMO, LAKE TURKANA, AND AWASH BASINS AND THE APPEARANCE OF THE HUMAN LINEAGE

Australopithecus anamensis • *Kenyanthropus platyops* • *Australopithecus afarensis* • *Paranthropus aethiopicus* • *Australopithecus garhi*

Volcanic activity is crucial for the preservation of fossils. When an animal dies, a veritable crew of organisms specialized as undertakers make use of all the energy still stored up in its tissues. Normally, in nature a dead animal's body lasts a very short time. Insects arrive immediately to lay their eggs in the flesh, hyenas and jackals to eat the flesh and crack open the bones, vultures and other birds to peck at the sinew, and rodents, beetles, and micro-organisms to consume the bone itself.

Volcanic ash may bury a carcass or living body immediately, sparing the bone from nature's undertakers and making fossilization possible. In some cases the hot ash-falls from volcanoes may bury a body at temperatures that carbonize the flesh but spare the bones. In this case, complete skeletons, free of breaks, insect tunneling, and tooth marks may be preserved, and the long process of fossilization can begin.

The African Great Rift Valley stretches from the Red Sea past the end of Lake Malawi, overlaying a rift or break in the

African continent. The rift is the result of the African continent pulling apart. It is associated with volcanic activity and both uplifting and sinking of land masses. The Great Rift Valley is marked by rivers that begin in mountain highlands on either edge of the rift and end in large land-locked basins at the rift's center, which are intermittently flooded. With no outlet to the ocean, the rivers deposit their sediments on the lowest lying land, leading to accumulations of massive deposits. Carcasses buried in ash in the floodplains of rivers are swept into channels and re-buried in deltas extending out into lakes or on the inside curves of growing river banks. The Awash River and the Omo and Lake Turkana basins, north and south of the Ethiopian plateau, are surrounded by volcanic activity. This geologic activity, during the millions of years human ancestors inhabited this area, has preserved a record of their evolution.

A MUCH WELCOMED VISIT

The morning rose with the songs of a thousand birds perched in the narrow strip of trees bordering the stream. It had been an uncharacteristically dry year. The March and April rains that finally came revived the trees, and their leaves grew with newfound strength. The birds had returned to feed on the insects that thrived with the new growth. Once again shade could be found below the trees that bordered the stream. Taking advantage of the cover, a lone man-ape walked along an old elephant trail, pausing to feed on herbs that sprung up in the moist low-lying patches of land flooded by the stream. Although the rains had revived the trees, it would be some time before the trees could produce the fruits so cherished by the man-ape. The man-ape, thus, satisfied her hunger with whatever greenery she could find. Bending back a young sapling and pulling off its leaves, she used her front teeth to strip off the bark and feed on the pith inside. By mid-afternoon she was joined by a dozen man-apes, sitting and feeding in a low-lying patch of land. Some ventured to the stream edge, pulling out grass shoots by the roots and feeding on the base of stems. Taking advantage of the late afternoon sun and a thick cloud cover, the man-apes ventured off into the scrubland, overturning rocks to look for insects and grub. Before nightfall they returned to the strip of trees bordering the stream to nest.

The following morning broke with a red sky and a stiff breeze from the north. The cloud cover increasingly thickened, screening out the sun, and stretching out the dawn. Looking towards the northeast, thin dark vertical streaks could be seen in the sky just above the horizon. A mid-morning light drizzle was followed by a subtle buzzing that grew in intensity as the sky darkened. At noon a large cloud of locusts descended on the land. With little else but foliage to eat, the locusts couldn't have come at a better time for the man-apes. A dozen or more man-apes sat in the open away from the narrow strip of trees gathering locusts with their hands, and stuffing them into their mouths. A francolin hen and her clutch fed right beside them. With eyes fixed on the ground and bringing handfuls of locusts up to their mouths, the man-apes fed for the remainder of the afternoon. As evening fell, a full moon, seemingly as large and orange-red as the sun, began its ascent from below the horizon. The man-apes returned to nest at the strip of trees by the stream. The following morning a baboon troop, a bushbuck, a marabou stork, and a hyena all fed beside the man-apes on the remaining locusts. Although the locusts had been a welcome addition to the man-apes' diet, they had left very little foliage in their wake. The strips of trees by the stream no longer provided shade. As long as the rains held up, however, it would take no more than a week before the foliage would return.

AUSTRALOPITHECUS ANAMENSIS

Skull, Teeth, and Diet

The most complete *Australopithecus anamensis* skull is represented by a lower jaw and a small fragment of the ear canal and jaw joint (KNM-KP 29281). An upper jaw (KNM-KP 29283) preserves both cheek tooth rows, a canine, and one front tooth. Other associated teeth from the Kenyan sites of Kanapoi (KNM-KP 29286 and KNM-KP 30500), including those of a child (KNM-KP 34725), and Allia Bay (KNM-ER 30745), have well preserved teeth and provide additional details.

The jaws are narrow from side to side with long parallel tooth rows resembling those of great apes. The cheek bones attach at the level of the first molar. The front teeth and canine are set well forward of the cheek bones and protrude from the face. Associated with the forward set of the lower teeth, the lower jaw has a strongly receding chin. The front upper teeth have beveled wear with strongly curved roots, which are set in a horizontally disposed bone plate narrowly separating the nose opening from the roof of the mouth. Together with the set of the cheek bones, the above features give *A. anamensis* an orangutan-like snout. The jaw joint surface is relatively wide from front to back and flat, showing a relatively mild curvature as seen in great apes. The ear canal has a relatively narrow diameter, which is chimpanzee-like and differs from the wider ear canal of later *Australopithecus* and *Homo*. Relative to the other teeth, both upper and lower canines are slightly larger than in other *Australopithecus* species and have large roots. Nevertheless, the canines show tip wear and do not appear to have been honed by each other or the lower premolars. Distinguishing *A. anamensis* from humans and later *Australopithecus*, its first lower premolar has only a single large cusp and the first milk molar is narrow with a large dominant cusp and a minimum surface area for crushing. Large variation in tooth size is possibly due to the presence of more than one species, or possibly differences between males and females.

Small molars with thick enamel suggest that *Australopithecus anamensis* probably ate fruits but also was able to handle an abrasive diet, which may have included seeds, leaves, bark, and other roughage. Protruding orangutan-like front teeth with beveled wear and a jaw joint with significant back to front movement reflect an emphasis on front tooth use for stripping and husking plants and fruits. As do orangutans, *A. anamensis* could probably secure a fruit with its lips and peel it with its front upper teeth. With the first milk molar unable to crush and grind, *A. anamensis* infants would not have eaten small grass seeds or other small plant items. Unfortunately, as of yet there are no *A. anamensis* tooth micro-wear studies that could reveal more about its diet.

Skeleton, Gait, and Posture

Two end fragments of a shin bone (KNM-KP 29285) have been claimed to show *Australopithecus anamensis* routinely moved about on two legs. According to its describers, evidence of this habit is provided by the right angle the bone's long axis makes with the ankle joint surface. Although some paleoanthropologists believe it provides substantial evidence of two-legged stance and movements, a similar ankle joint orientation is seen in baboons and other four-legged ground-moving Old World monkeys. Additional evidence is thus needed to demonstrate conclusively that *A. anamensis* stood and moved around on two legs.

The forearm (KNM-ER 20419) and finger bone (KNM-KP 30503) suggest that *Australopithecus anamensis* would have used its upper limbs for support when moving about in trees and on the ground. The forearm bone is relatively long (265 mm to 277 mm) with a mid-shaft cross-section to length proportion that is chimpanzee-like. In its curvature, joint orientations, and muscle, tendon, and ligament attachment sites, it is most similar to an orangutan. The chimpanzee-like curvature of the finger bones indicates the tendons controlling finger movement applied the greatest force with flexed finger joints, as occurs when an ape hangs in a tree or knuckle-walks.

On one hand, the describers of *Australopithecus anamensis* have suggested these features were inherited from a primitive ancestor. On the other hand, shaft diameters and curvatures and joint orientations are known to change during an individual's lifetime according to behaviors practiced. So whether these features were inherited or the result of lifetime usage is an unanswered question.

Since neither the limb nor finger bones were associated with the *Australopithecus anamensis* dental and cranial fossils, it is uncertain if they belong to *A. anamensis*. Because they span a time interval of 300,000 years and could very well belong to different species if not genera, the use of these fossils to reconstruct *A. anamensis* behaviors is speculative. Likewise, a 47 kg to 55 kg weight-estimate, based on the shin bone and assuming human body dimensions, cannot be guaranteed accurate.

Fossil Sites and Possible Range

Australopithecus anamensis is found in the upper part of the Lonyumun Member of the Koobi Fora Formation at Allia Bay on the east shore of Lake Turkana and in sediments above and below the Kanapoi Tuft at Kanapoi, southwest of Lake Turkana, Kenya. The Kanapoi sediments yielding *A. anamensis* are largely sandstones deposited by the prehistoric Kerio River channel and its delta as it entered into Lake Lonyumun. Lake Lonyumun was the precursor of Lake Turkana, covering an area about three times as large. The upper part of the Lonyumun Member at Allia Bay yielding *A. anamensis* is mainly a sandstone channel deposit from the ancestral Omo River. Animal fossils scattered in the river's floodplain were swept into the faster channel waters and were concentrated. Many Allia Bay fossils are water-rolled fragments that resisted breakage, mostly teeth. Kanapoi fossils are not as fragmented as the Allia Bay ones, but all seem to have been worked over by hyenas or other carnivores.

Australopithecus anamensis was also reported from Aramis in the Andangatole Member of the Sangatole Formation and 10 km west of Aramis at Asa Issie and Hani Hara, Middle Awash River, Ethiopia, in unnamed sediments more or less contemporaneous with the Andangatole Member. Reported fossils, mostly teeth, are too incomplete to confidently assign to *A. anamensis*. Fejej, a northeast Omo Basin site in Ethiopia from approximately the same age interval as Kanapoi and Allia Bay, yielded some isolated teeth initially assigned to *Australopithecus afarensis*, which given their age are more likely to be *A. anamensis*. These fossils, however, are also too incomplete to classify with certainty.

Age

Basalt flows and layers of volcanic ash (tuffs) above and below the Allia Bay and Kanapoi deposits yield reliable absolute dates for all the fossils. The Lonyumun Member is sandwiched between a basalt flow and the Moiti Tuff, yielding radiometric dates of 4.35 million years and 3.89 million years. Assuming a constant deposition rate, each meter of the thirty-seven-meter-thick member is calculated to have taken 12,432 years to deposit. Found 5 m below the Moiti Tuff, most Allia Bay

Australopithecus anamensis

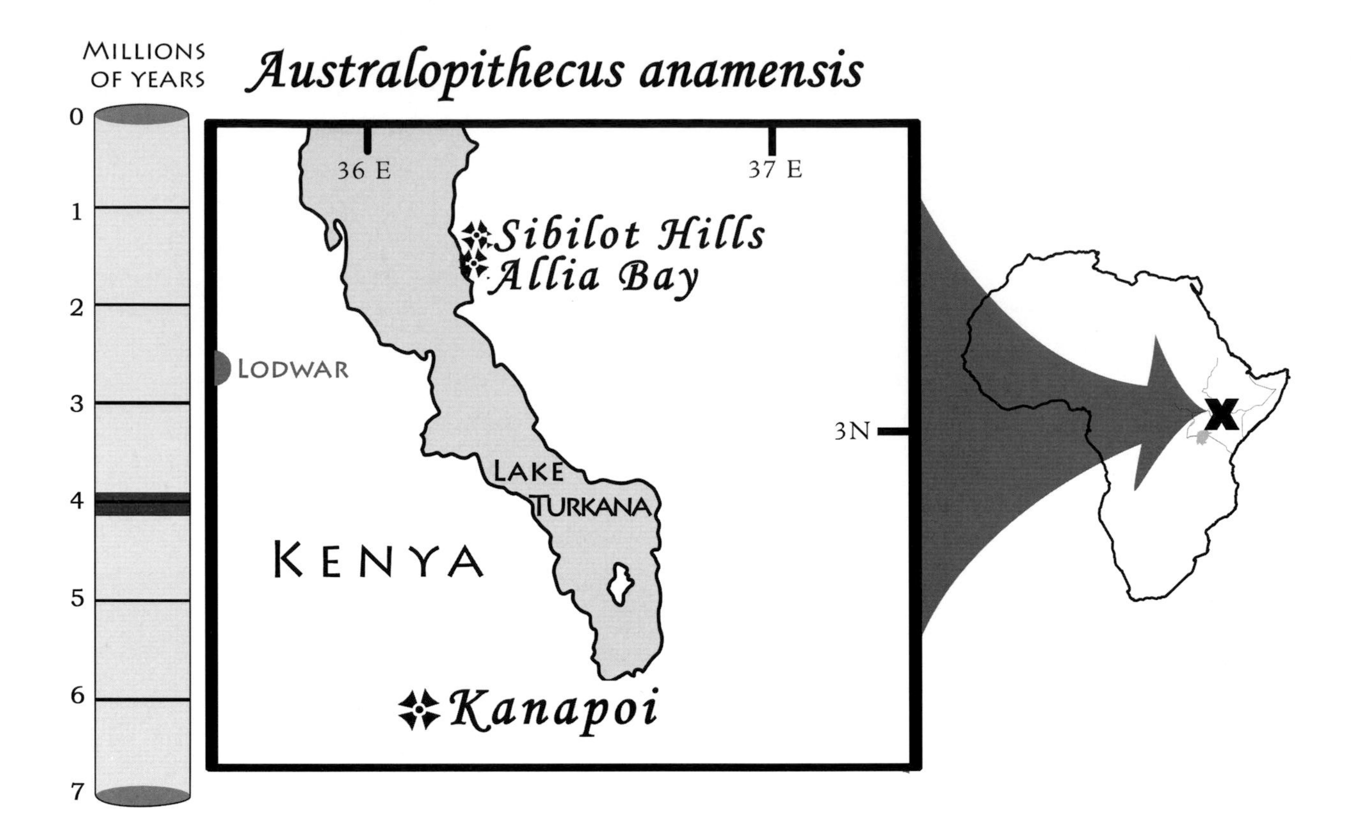

Australopithecus anamensis fossils have an age of 3.95 million years. The isolated forearm bone (KNM-ER 20419), which was found in Sibilot Hills, 20 km northeast of the other Allia Bay fossils, is about three million nine hundred thousand years old. Radioisotope dating of the Kanapoi Tuff and of two underlying pumice tuffs yielded dates of 4.07 million, 4.12 million, and 4.17 million years.

All *Australopithecus anamensis* fossils are found below the Kanapoi Tuff, with the exception of a lower jaw (KNM-KP 29287) that comes from above. This jaw is therefore younger than 4.07 million years. Two fossils, a shin bone (KNM-KP 29285) and a lower arm bone fragment (KNM-KP 271), were found between the Kanapoi Tuff and the first underlying pumice tuff and are thus 4.07 million to 4.12 million years of age. The remaining fossils were found between the underlying pumice tuffs and thus have an age between 4.12 million and 4.17 million years.

Fossils from Aramis assigned to *Australopithecus anamensis* come from deposits between layers of volcanic ash (tuffs) with radiometric dates of 4.32 million and 4.04 million years. Sediments in which the fossils have been found show Earth at this time had a reversed magnetic field, limiting these fossils to a maximum age of 4.21 million years. They are thus closer to Kanapoi *A. anamensis* in age. Assa Issie and Hani Hara *A. anamensis* are found between layers of volcanic ash with radioisotope dates of 4.17 million and 3.77 million years. It appears these fossils are closer to the older of these two dates, but this is by no means certain.

Tools

Australopithecus anamensis predates the appearance of tools by as much as one million four hundred thousand years.

Appearance

The orangutan-like horizontal orientation of the plate of bone holding the roots of the upper teeth suggests an orangutan-like nose, which was positioned substantially behind the upper lip and front teeth. Aside from the links between a tropical sun, high radiation, and dark skin, with just a jaw and no other parts of the skull, most of its appearance is conjectural.

Animals and Habitats

Many animal genera found with *Australopithecus anamensis* at Kanapoi and Allia Bay are now extinct. At Kanapoi these include a prehistoric baboon (*Parapapio*), an undescribed leaf-eating monkey, a large otter (*Enhydrodon*), a precursor of the brown hyena (*Parahyeana*), a three-toed (*Parahyena*), horse (*Hipparion*), two species of a bush pig (*Nyanzachoerus*), an early relative of gnus and wildebeests (*Damalacra*), two giraffe species, and two prehistoric elephants (*Deinotherium* and *Anancus*). From Kanapoi a relative of cattle and a relative of roan and sable antelopes have yet to be identified and may also represent extinct genera. Allia Bay also yielded two extinct large cats not found at Kanapoi, *Dinofelis* and an unidentified saber-tooth. Modern mammal genera from Kanapoi include a giant elephant shrew, a bush baby, a gerbil, a porcupine, an African elephant, a white rhinoceros, a pygmy hippopotamus relative, two extinct African giraffe species, a bushbuck or nyala, a kob, a gazelle, an impala, a dik-dik, and a steenbok. Allia Bay also yielded a species of Indian elephant, a shrew, a hedgehog, a very small monkey, a multi-mammate mouse, and an acacia rat. A member of the rat family and a squirrel resisted identification. At both sites all animals that could be identified to species belonged to species that today are extinct.

Although ancient rivers emptying into Kanapoi and Allia Bay cut through many habitats, washing away mixed animal remains, the fossils of woodland animals are the most prevalent at both these sites. At Allia Bay, the large number of bushbuck/nyalas and the presence of bush babies, acacia rats, multi-mammate mice, and a small monkey suggest a mosaic of woodlands with grassy clearings and small secondary forest patches close to water sources. There are no known forest animals, so it is unlikely that *Australopithecus anamensis* would have been one. Its narrow ear canal, however, suggests that it did not evolve in open habitats devoid of trees. Animals in open habitats usually have wide ear canals sensitive to low-frequency sounds, which can move long distances when unobstructed by trees and foliage.

Climate

Weather records from Lodwar (506 m above sea level), a town on the southwest side of Lake Turkana, Kenya (120 km northwest of Kanapoi and 90 km southeast of Allia Bay), show temperatures vary very little during the year. Maximum monthly averages range from 33° C (91.4° F) in July and August to 37° C (98.6° F) in February. Minimum monthly averages range between 22° C (71.6° F) in January and 25° C (77° F) in May. At an elevation 130 m lower, the lake and the river delta would enjoy temperatures nearly 1° C higher.

Average annual precipitation is 165 mm, with an annual maximum (over a forty-four-year period) of nearly 500 mm. More than half the annual rainfall occurs during March-May, but in very wet years a second peak may occur from November-January. Annual rainfall amount increases northwards. Lake Turkana's northernmost part receives a mean of 500 mm of rainfall annually, and the northern extent of the Shungura Formation, 60 km to the north of the lake, may receive as much as 1000 mm annually. There is no evidence that the climate would have been much different today than it was during *Australopithecus anamensis* times. The rainfall would thus have been too low to support forest conditions.

Classification

The large range in size and also appearance exhibited by the *Australopithecus anamensis* molars has prompted some paleoanthropologists to postulate as many as five different early hominids just within Kanapoi. Although five seems to be excessive, Kanapoi shows a range of variation that is too great to fit in a single species. When the Allia Bay

A. anamensis fossils are included with those of Kanapoi, the presence of two or more species becomes all the more obvious.

The Ethiopian Aramis, Asa Isse, and Hani Hara fossils are too incomplete and too worn and damaged to allow certain classification. Adding to this difficulty is the problem that the most complete Ethiopian fossils are not directly comparable to the *Australopithecus anamensis* holotype lower jaw specimen. Moreover, they do not show features that are exclusive to *A. anamensis*, so that only size similarities and the similar age of deposits in which they were found can be used to justify placing them in *A. anamensis*. More complete finds will most likely show the Ethiopian fossils belong to their own species. Seven degrees' difference in latitude (800 km) between Allia Bay and the Ethiopian fossil sites supports this likelihood.

Given its chimpanzee-like premolar, milk molar, and narrow ear canal, all features it shares with *Ardipithecus*, it is not clear why *Australopithecus anamensis* is classified in the genus *Australopithecus*. These features, and its orangutan-like snout, suggest to some paleoanthropologists that *A. anamensis* may predate the human–African ape split. With uncertain evidence as to its two-legged movements, its definitive assignment to the hominid lineage is in need of more supporting evidence.

Historical Notes

Bryan Patterson is credited with finding the first remains of *Australopithecus anamensis*, a lower arm bone (KNP 271) at Kanapoi in 1965. Patterson and William Howells reported this find in a 1967 issue of *Science*. Other than to point out its human-like features, they determined it too fragmentary to classify its species or genus. The Lake Turkana Allia Bay site, where most *A. anamensis* fossils come from, was discovered in 1968. Jonas Kithumbi found the first Allia Bay *A. anamensis* fossil, an isolated upper molar (KNM-ER 7727), in 1982.

In a 1994 issue of the *American Journal of Physical Anthropology*, Katherine Coffing, Craig Feibel, Meave Leakey, and Alan Walker assigned the Allia Bay fossils tentatively to *Australopithecus afarensis*, but realized their hominid status was uncertain. A year later in *Nature*, Leakey, Feibel, McDougall, and Walker, reporting on the shin bone, described the Allia Bay and Kanapoi fossils as a hominid and referred then to *Australopithecus anamensis*, choosing the KNM-KP 29281 lower jaw and small fragment of the ear canal and jaw joint from Kanapoi as the holotype specimen. Ian Tattersall, in Volume 4 of *The Human Fossil Record* (Wiley-Liss 2005), postulated that the Kanapoi *A. anamensis* fossils represent at least five different types. In April 2006, Tim White and colleagues found *A. anamensis* fossils in the Middle Awash River deposits in Ethiopia, declaring this species to be the evolutionary link between *Ardipithecus* and *Australopithecus*.

The species name "*anamensis*" comes from the word "aman," which means "lake" in the Turkana language. It was chosen because of the fossil's association with lakes (i.e., Turkana and Lonyumun). Kanapoi and Allia Bay fossils are housed in the Kenya National Museums, Nairobi. Aramis, Assa Issie, Fejej, and Hani Hara fossils are housed in the National Museum of Ethiopia, Addis Ababa.

A troop of *Australopithecus anamensis* cautiously makes its way through the forest in search of water and shelter for the night.

WE ARE FAMILY

The mid-day sun warmed the air, kicking up dust devils across the scrubland. Hit by the whirling dust, a herd of topis took flight. A giraffe browsing on the leaves of an acacia tree and a gelada baboon troop had to move quickly to avoid being trampled. Running in a huge circle, the topi herd finally came to rest near a group of man-apes feeding in and around several large bush-mango trees on the edge of the woodland lining the lakeshore. Although still green, the bush-mango fruit had already reached its maximum size and was there for the man-apes' taking. As soon as the topi herd arrived, the man-ape closest to the herd became animated, shaking her head and swatting at the air around her. The tse-tse flies that followed the topis were close enough to the man-apes to try their luck with a different host. Ultimately the flies put an end to the man-apes' lunch, driving them back through an old elephant trail into the woodlands.

The man-ape group displaced by the tse-tse flies was relatively large and soon to become larger. At its head was an old male. With a bald pate and silver hair peppered black on his back and shoulders, his movements were slow and methodical. A scar that ran across his nose and upper lip, and continued on the right side of his chest, was witness to a crocodile attack survived as a youngster. The river bank had given way and dropped the man-ape into a crocodile's waiting jaws. Screaming, he managed to escape onto shore, beating the crocodile back with a branch.

The old man-ape had had a number of mates, but only two now survived. The oldest of his mates was a thin long-limbed female with jet black hair. She had given birth to two males that were still in the group, her first born, an adult, and fourth born, an eight year old. She was now nursing a two year old daughter. Given the mother's age, this infant was probably the last offspring to which she would give birth. Her second and third born, male and female twins, had left the group two years before to find mates.

The second of the old male's mates had a five year old son and was pregnant with a second child. Larger and much younger than the other mate, she had a playful disposition and was quick to interact with all the youngsters in the group. A distinctive white lock of hair that fell on the right side of her forehead made it easy for the others to recognize her at a distance.

The first born son of the old man-ape and his first mate was a very large adult male. Like his mother, his coat color was jet black, although a grey hair here and there could be seen when close up. Like his mother, he was also long-limbed, though larger and more muscular. He was very excitable, with a moody disposition. Upon leaving his nest in the morning, he would often engage in violent displays, uprooting plants, breaking tree limbs, and running over other man-apes as he called out at the top of his lungs.

The first born son had two mates. The oldest was the mother of a four year old daughter, and was now pregnant awaiting a second child. She had a shiny dark brown coat and hazel colored eyes. She was the largest of the females in the group, with a quiet and self-assured manner. The other mate was a very young female who had yet to finish growing. Nonetheless, she was pregnant. In contrast to the other mate, she was nervous and insecure and would shriek and cower at the displays of the first born male.

Signaling they were pregnant or lactating, all the breeding age females in the group had enlarged breasts. The two adult males showed considerable deference to them, when both feeding and choosing areas in which to nest. The appearance of the expectant mothers had totally changed the first born son's behavior. In the last month he had displayed only twice, and seemed to be more at ease and less anxious, appearing overall more content.

A long-legged female with long brown hair would occasionally join the group. Two days ago she had been calling and presenting herself to the first born son. Without a mate she had been moving among the man-ape groups in the area. Something about this group made her feel comfortable. She now joined the group around a persimmon tree feeding on the fruit. The males did not pay much attention to her, but the old male's youngest mate fed nearby and appeared to accept her. If the long-legged female joined the group and the three pregnancies were carried to term, the group's size would increase from nine to thirteen.

KENYANTHROPUS PLATYOPS

Skull, Teeth, and Diet

A distorted and badly abraded skull (KNM-WT 40000) and a fragmentary upper jaw (KNM-WT 38350) are all the fossils that can be clearly assigned to this species. One lower jaw with all of the right cheek teeth (KNM-WT 8556), a small fragment of the skull with an ear canal (KNM-WT 40001), and thirty-three other fossils, mostly isolated teeth and upper and lower jaw fragments, may also belong to *Kenyanthropus*, but lack features unique to *Kenyanthropus* that can definitively associate them to the skull or upper jaw. Three of these fossils are from the same Kenyan deposit as the skull and have the same approximate age. The other remains are from the same deposits as the jaw (KNM-WT 38350) and are close in age or younger than the jaw.

The skull (KNM-WT 40000) preserves only the incomplete crown of the second molar, and thus molar dimension can only be estimated. The ear canal is long and narrow and the ear opening is small, as it is in *Australopithecus anamensis* and living chimpanzees. The skull's most distinctive feature is the forward position of its very large and flat cheek bones relative to the jaws. There is a rather broad, flat, and nearly upright-set bone plate below the nose opening holding the straight roots of the front teeth, and broadly separating the nose opening from the mouth and palate. Together with the large forward-set cheek bones, this plate gives the face a flat-dished appearance. The single first and second molars of the KNM-WT 38350 fragmentary jaw and KNM-WT 40000 skull appear to be small with thick enamel, but within the size range of *A. anamensis*. The cracked and considerably distorted palate is broad and shallow. As in robust australopithecines and some *Homo* species, the upper premolars have three roots instead of two.

The skull is too damaged to measure brain volume, but given approximate skull breadth and length, it would not have exceeded 400 cc. A presumed male status for this skull is based on the slightly raised and closely placed attachment ridges for the chewing muscles on either side of the midline along the top of the braincase. With only a sample of one, however, the sex of the skull cannot be determined. Unfortunately, the skull base was not preserved, so the position of the foramen magnum is unknown.

Studies on the *Kenyanthropus* diet have yet to be undertaken. Absence of a complete dentition or even unworn molars severely limits most speculation. Like great apes, it must have eaten herbs, succulents, and leaves. Its relatively small molar size suggests it may have depended on fruits. Living well within the tropics in what appears to be a well-watered area, *Kenyanthropus* would have eaten fruits year round.

Skeleton, Gait, and Posture

There are no known skeletal bones for *Kenyanthropus*. A misidentified finger bone (KNM-WT 16004) from the lower Lomekwi Member is neither ape nor human. A right fragmentary thigh bone missing the joint surfaces (KNM-WT 16002) is

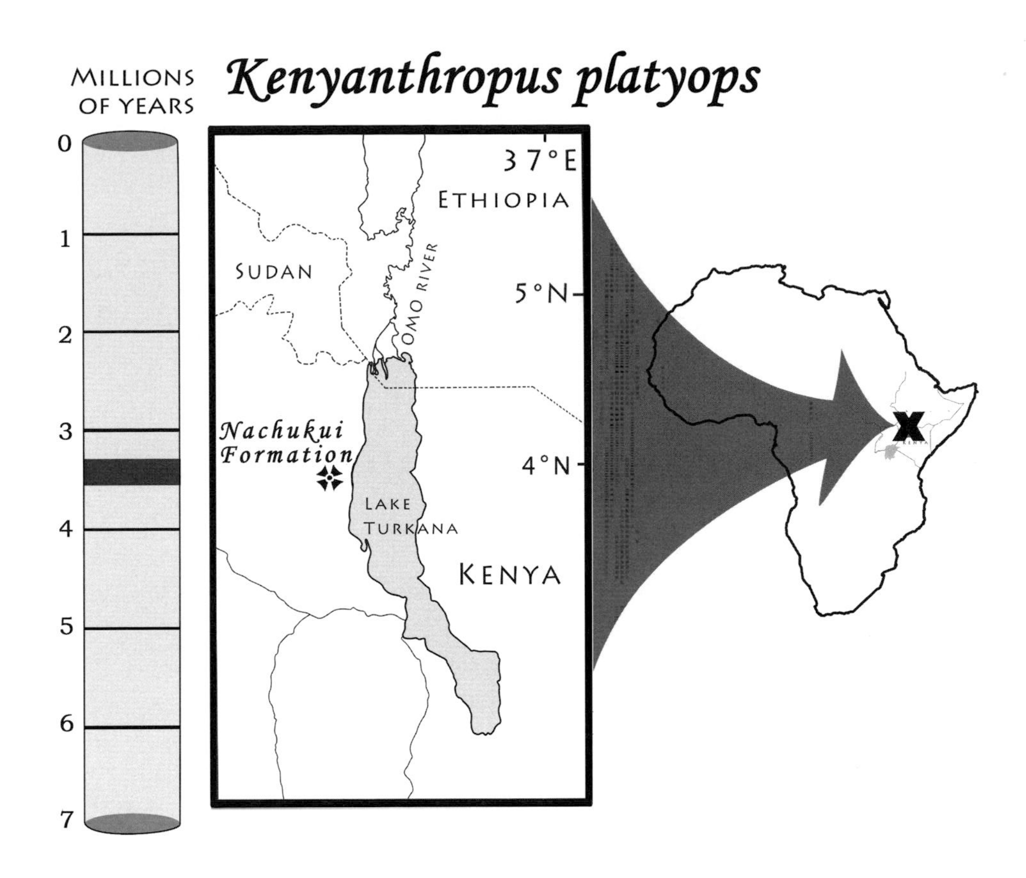

from a higher part of the Lomekwi Member and dates to between 3.2 million and 3.3 million years. Although this bone could conceivably belong to *Kenyanthropus* or its descendants, it provides no clear evidence as to how this animal moved when it was alive.

Fossil Sites and Possible Range

Kenyanthropus platyops fossils are found in mudstones, sandstones, and conglomerates of the lower Lomekwi and Kataboi Members of the Nachukui Formation on the northwest side of Lake Turkana, in and around the drainage areas of the Lomekwi and Topernawi Rivers, Kenya. Mudstones, sandstones, and conglomerates accumulated from deposition 1) along the margin of a shallow lake, 2) in floodplains, and 3) in short-lived streams. Considering their proximity and a similar time interval, non-identified early hominid fossils from the Usno Formation, from Members B through B3 of the Shungura Formation (Ethiopia) and from the Tulu Bor Member of the Koobi Fora Formation may also belong to *Kenyanthropus*.

Age

The Kataboi and Lomekwi Members are sandwiched between layers of volcanic ash, which yielded precise radioisotope and paleomagnetic dates. Found between the Lokochot and Tulu Bor tuffs and between the Tulu Bor and Loklalei tuffs, the Kataboi and Lomweki Members span 3.4 million to 3.57 million years and 2.5 million to 3.4 million years, respectively. Assuming a constant deposition rate over 0.17 million years, each meter of the twenty meter-thick Kataboi Member took 8500 years to deposit. Found at 8 m below the Tulu Bor tuff, the skull is somewhat less than three and a half million years old. The upper jaw (KNM-WT 38350) was found 17 m above the Tulu Bor in a member which is 150 m thick and is thus estimated to be about three million three hundred thousand years old.

Tools

There are no tools associated with *Kenyanthropus platyops*. Its fossils predate the oldest known tools by at least seven hundred thousand years.

Animals and Habitats

Unfortunately, there are no known small mammals in the *Kenyanthropus platyops* bearing deposits, making habitat reconstructions difficult. All mammals found have a relatively large body size and represent a mixed bag of extinct and living genera. Extinct genera include a prehistoric baboon (*Parapapio*), the scimitar cat (*Homotherium*), a prehistoric elephant with lower jaw tusks (*Deinotherium*), a three-toed horse (*Hipparion*), a short-neck giraffe (*Sivatherium*), four pigs (*Nyanzachoerus, Notochoerus, Kolpochoerus*, and *Metridiochoerus*), and a relative of hartebeests and gnus (*Parmularius*). Separated into a maximum of three groups based on body size differences (i.e., big, medium, and small), three monkey groups, two reedbuck/kob groups, three hartebeest/gnu groups, one group of cattle/bison, and one group of roan/sable have resisted identification and may also contain extinct genera.

Modern genera are represented by a gelada baboon (*Theropithecus*), the laughing hyena (*Crocuta*), two African elephant species (*Loxodonta*), one Indian elephant species (*Elephas*), a bush pig (*Potamochoerus*), both pygmy and river hippos (*Hexaprotodan* and *Hippopotamus*), an unidentified giraffe species (*Giraffa*), three bushbuck or nyala species (*Tragelaphus*), both black and white rhinoceroses (*Diceros* and *Ceratotherium*), two waterbuck or kob species (*Kobus*), a topi (*Damaliscus*), an impala (*Aepyceros*), two gazelle species (*Gazella*), and a springbok (*Antidorcas*). Considering the range of early hominid tooth sizes in the *Kenyanthropus* deposits, it is likely that it also shared its habitat with another as of yet unidentified great ape and/or hominid species.

The occurrence in the same deposits of closely related animals, i.e., four different elephant species, five different pig genera, five monkeys, both living genera of hippos, and both living genera of rhinos, all of which must have occupied different habitats, indicates a water source visited by animals at the interface of scrubland, bushland, woodland, and grass covered wetlands, or changes in habitats over the time interval the deposits accumulated. There are no forest animals or ones that are associated only with forest edge. More than likely *Kenyanthropus* shared woodlands along streams and lakeshores with monkeys and other primates.

Climate

(*See Australopithecus anamensis.*)

Classification

The forward set of the cheek bones relative to the tooth row, and the set, shape, and dimensions of the bony plate holding the upper front tooth roots, distinguish *Kenyanthropus platyops* from all *Australopithecus* species and justify placing it in a new species. Reasons as to why it merits its own genus, however, are not so evident. Because it shares some features with *Australopithecus anamensis* (i.e., narrow ear canal and small molars), which suggest the two have a common relationship exclusive of *Australopithecus africanus* and *Australopithecus afarensis*, there is no justification for *K. platyops* being assigned its own genus while *A. anamensis* remains in *Australopithecus*. Moreover, if, as argued by the describers, *K. platyops* shares unique features (associated with face shape) with the two-million-year-old *Paranthropus* and *Homo rudolfensis*, which distinguish them from all *Australopithecus* species, its classification in the genus *Homo* or *Paranthropus* must be further

Caught off guard from more pressing duties, a startled male *Kenyanthropus platyops* springs to his feet from behind a thick wall of vegetation. A competing group of *Australopithecus afarensis* have made their way into his feeding grounds.

explored. Given what is known of *Kenyanthropus*, placing it in a separate genus appears to require a revision of hominid classification.

Kenyanthropus platyops presents a few features (all of which are associated with front tooth and canine reduction) that suggest it is a member of the human lineage, post African ape divergence. While many paleoanthropologists have accepted these features as showing proof of its membership in the human lineage, more skeptical ones would argue that additional fossil material is necessary, especially of the feet and lower limbs, to show this is the case.

Historical Notes

B. Onyango is credited with finding the first identifiable fossil of *Kenyanthropus*, a fragmentary maxilla (KNM-WT 38350), in 1998. J. Erus found the skull a year later. Based on these, Meave Leakey, Fred Spoor, Frank Brown, Patrick Gathogo, Christopher Kiarle, Louise Leakey, and Ian McDougall coined the binomial *Kenyanthropus platyops* in a March 2001 issue of *Nature*. In a March 2003 journal of *Science*, Tim White argued that the *Kenyanthropus* skull had suffered way too much distortion to serve as the basis for a new species and could be a variant of *Australopithecus afarensis*.

The generic name "*Kenyanthropus*" means "man from Kenya" and was chosen in recognition of Kenya's contribution to understanding human evolution. The species name is from Greek "platus," meaning "flat," and "opsus," meaning "face," and literally translates into "flat face." The fossils are housed in the National Museums of Kenya, Nairobi, Kenya.

THE FIRST NOMAD?

AUSTRALOPITHECUS AFARENSIS

Each year as a result of the summer rains the swollen river would overflow its banks, and water the surrounding land. The cattails would grow high on the mudflats along the flooded river banks and leaves would sprout on the gum myrrh trees. In time, the trees growing in the forests flanking the river would bear fruit and the ground man-apes would return from the highlands to feed.

Standing in the crook of a tree, a female man-ape reached up for unripe figs. Leaf monkeys jumped back and forth in the smaller branches of the tree crown dropping partly eaten figs on the man-ape below. One leaf monkey descended down to the man-ape's eye-level. Facing the man-ape, it chattered and squealed at her relentlessly. Harassed by the noise and debris, the man-ape descended the tree feet first. Remaining on two legs, she leisurely walked to another tree and picked the fruit from the lower branches. The leaf monkey followed, coming down to continue its banter at what it must have assumed to be a safe distance. Bounding from two legs to four, the man-ape grabbed hold of both of the monkey's hindlimbs. Placing her free hand on the monkey's neck and bearing down with her full weight, she quickly broke its back, silencing its squeals. Retiring behind some shrubbery, the man-ape hid to enjoy her meal in private, but it was too late. Other man-apes in the group had heard the commotion. Hooting and hollering, man-apes came out of nowhere to beg a meal. In a few minutes, the entire group was there. The monkey was pulled apart and its parts passed from hand to hand. With the bones bared, the carcass was taken by the youngsters, who pulled at it and bit at the sinew surrounding the joints that were still intact.

All the noise that had ensued with the distribution of the monkey meat came to the attention of another man-ape group that was feeding on herbs along the river bank some distance downstream. A dominant male in this other group had been calling ever since, and the calls now seemed to be getting closer. In time, an old male appeared from behind the undergrowth. After him came five females, two young males and an assortment of youngsters.

Standing on two legs, the dominant males from each group came face to face. When they reached arm's length, the old male dropped down to a squatting position. The other man-ape acknowledged this gesture by bending over and stroking the hair on the old man-ape's head and shoulders. Man-apes from both groups now came together, greeting and grooming each other amid grunts and chatters. The youngest members of the groups wrestled and ran after each other on all fours. Jumping into shrubs or small trees, they would end the chase only to have it start again in the opposite direction, with the ones being chased now doing the chasing.

Emerging from the thick underbrush of the riverine forest, the man-apes assumed a two-legged stance to feed on pods and seeds in a grove of acacia trees. With the late afternoon sun starting its descent below the horizon, some of the man-apes constructed nests in the acacia trees. Others, mostly the larger males, built their nest on the ground at the base of trees from grasses and herbs. In time, the extra food that came with the summer rains would be gone, and the man-apes would go back to the evergreen forest in the highlands. Although not as productive in terms of food as the lowlands, the higher rainfall in the highlands insured the growth of greenery through out the year.

AUSTRALOPITHECUS AFARENSIS

Skull, Teeth, and Diet

The Hadar, Ethiopia, deposits have produced a wealth of *Australopithecus afarensis* skull fragments and teeth, the most complete being the A.L. 822-1 skull. Unfortunately, this skull has yet to be reported on. For the moment another skull (A.L. 444-2), from the lower Kada Hadar Member approximately three million years old and almost as complete as A.L. 822-1, forms the basis of what we know of *Australopithecus afarensis*. The skull is large, with a very large face that is hafted low relative to the braincase. The cheek bones are very wide from top to bottom and side to side, projecting out to the sides considerably past the eye sockets. The cheek bones attach relatively low on the upper jaw at the level of the first molar. Despite the width of the face, the eye sockets are relatively close to each other and the upper nose narrow. The nose opening is elongated and tear-drop shaped. Although its upper border has not been preserved, it appears to be at the level of the eye sockets lower border. The plate of bone in which the upper front tooth roots are set is horizontally disposed. The front teeth curve downwards and have beveled wear. The nose opening is narrowly separated from the mouth. Owing to a forward set of the tooth rows, the upper jaw juts out past the face, creating a noticeable orangutan-like snout. The skull lacks a forehead, and the braincase behind the eye sockets is narrow from side to side. The low attachment of the neck muscles on the back of the skull gives the braincase a much higher appearance than in living chimpanzees and gorillas. From front to back, the skull base is short. But, it is very wide from side to side, corresponding to the width of the face. The skull base faces almost directly downwards. In its orientation and in a number of other characteristics, the skull base is very similar to that of *Paranthropus*. At 550 cc, the A.L. 444-2 braincase volume is the largest of the five *Australopithecus afarensis* skulls that can be measured and considerably above the 446 cc average (range 550 cc to 387 cc).

The cheek tooth rows are long, with the right and left rows set parallel and close to one another. Despite the great width of the face, the palate is narrow. The chewing surface of the upper teeth curves upwards from back to front so that the plane of the chewing surface is much higher at the front teeth than at the molars. The jaw joint is narrow from front to back, limiting back and forth movement between the upper and lower jaws. Despite having very similar molar and premolar cusp patterns, the cheek teeth are smaller than those of *Australopithecus africanus*. (*see Australopithecus africanus* Skull, Teeth, and Diet)

Smaller cheek teeth than *Australopithecus africanus* suggest *Australopithecus afarensis* would have consumed a more nutritious diet with less bulk and roughage than *A. africanus*. A narrow palate relative to a wide face is associated with a balanced distribution of the right and left jaw muscle force on the chewing surface, so that during chewing muscles either side can apply a nearly maximum force on either the left or right tooth row. The upward curved chewing surface of the

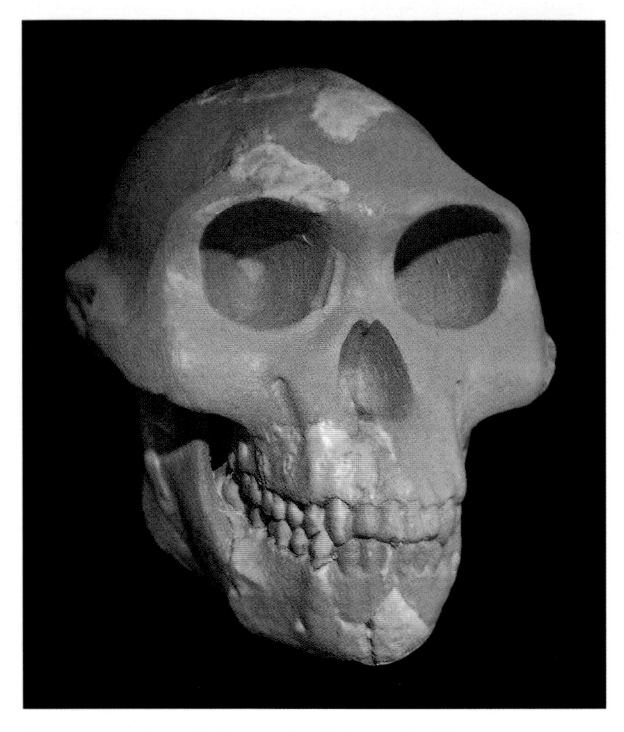

Reconstruction of the skull of *Australopithecus afarensis* (Lucy), revealing a small jaw and reduced canines.

cheek tooth row serves to distribute muscular forces more evenly between the back and front cheek teeth. In mammals, all these mechanical advantages are associated with hard object feeding (nuts and seeds), and to one degree or another are also seen in orangutans. As in orangutans, the protruding upper front teeth of *A. afarensis* would have been important in food processing, even much more than in *A. africanus.* Front tooth micro-wear analysis shows *A. afarensis* used the front teeth to strip gritty plant foods. Overall the *A. afarensis* specializations suggest a fruit-eater that would have consumed fruits and leaves in nearly equal amounts. Pith, seeds and herbs would also have been important items in the diet.

Skeleton, Gait, and Posture

The *Australopithecus afarensis* skeleton is well represented by the A.L. 288-1 skeleton (nicknamed Lucy); hand, foot, and limb bones from the excavated 333 locality; and most recently by the Dikika infant skeleton (DIK-1-1). Many bones are complete and preserve joint surfaces, affording insights into locomotor behaviors. Based on long bone shaft cross-sections, Lucy's estimated body weight is between 26 kg and 29 kg, considerably below the average for all great ape species, but closest to living pygmy chimpanzees.

Some paleontologists have interpreted traits in the *Australopithecus afarensis* skeleton as adaptations to two-legged ground movement. Principally, a more human-like than chimpanzee-like pelvis in Lucy and the presence of a carrying angle (*see Homo sapiens*) in some incomplete thigh bones (A.L. 129-1a, A.L. 333-4, and A.L. 333w-56) have been interpreted as evidence that *A. afarensis* moved around on two legs. These traits, however, are also seen in climbing animals and are not exclusive to two-legged behaviors.

Bone thickening in the lower half of the thigh bone neck shaft (MAK-VP-1/1) and upward facing joint surfaces in the toe bones (A.L. 288-1, A.L. 333w, and KNM-WT 22944 K) have also been cited as evidence of two-legged movements. But these traits are also seen in baboons and other four-legged ground moving animals. Because none of these features are exclusive to a two-legged stance, they do not provide conclusive proof that *Australopithecus afarensis* was a habitual biped.

Relative to arm bone, thigh bone, and finger and toe bone lengths, Lucy's body size is most similar to living baboons. As in baboons, its arm bone to thigh bone ratio is 85 percent. Its relatively short fingers and toes indicate it had sacrificed tree-climbing for ground movement. Climbing apes need long arms (more than 90 percent arm to thigh bone length ratios) for reaching, and long fingers and toes for grasping. Lucy is similar to humans and gorillas in having a very large ratio of finger to toe bone length. Its short toes indicate that like humans and gorillas it could not have hung from its feet. Most of its climbing would have been on top of large branches or tree trunks.

Australopithecus afarensis

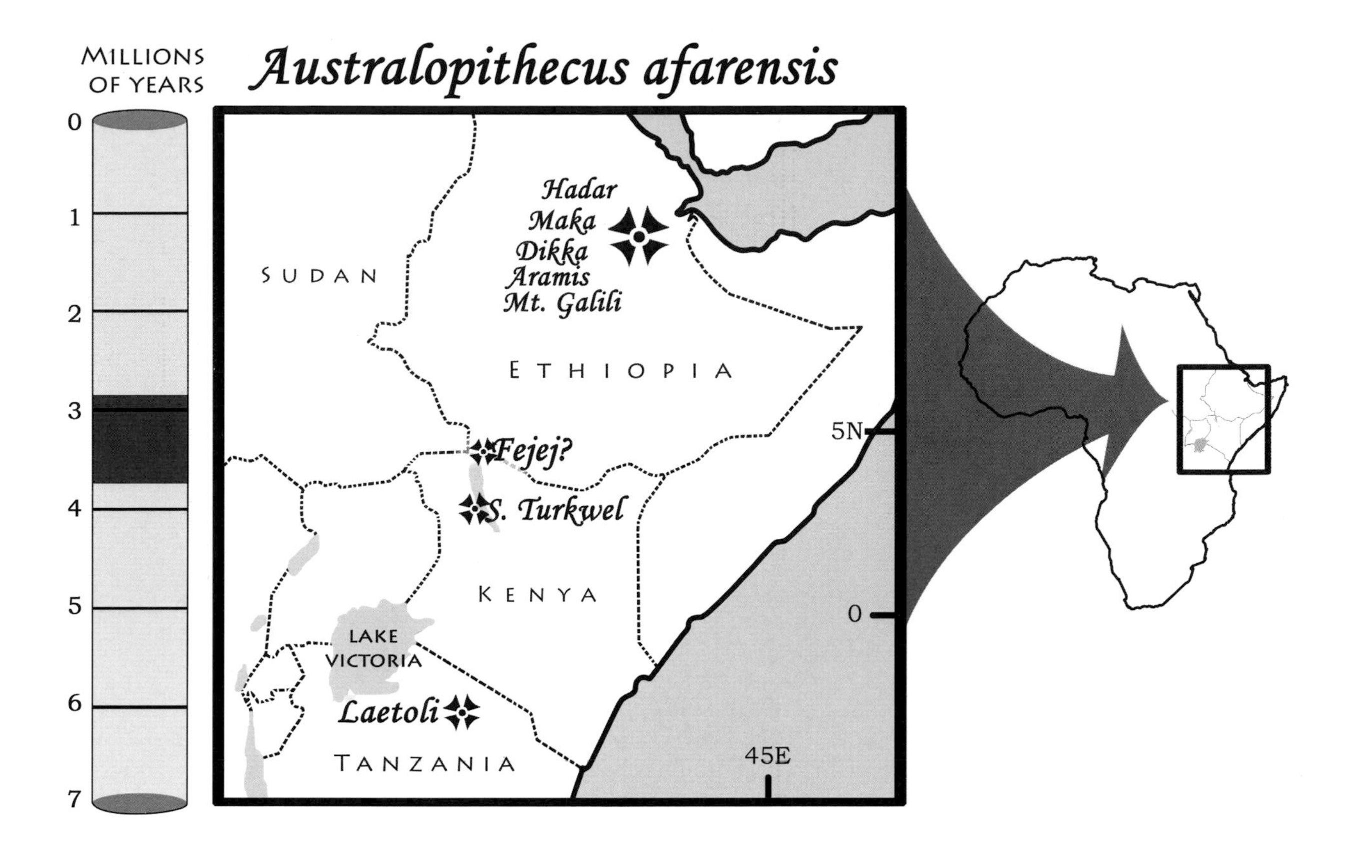

Relative to body weight, Lucy's arm bone shaft has a large cross-section, while the joints between the lower back bones have relatively small surface areas, suggesting the lower back would not have customarily supported the trunk weight without help from the forelimbs. Lower back bone joint surfaces and long bone shaft cross-sections are well known to change during an individual's lifetime in response to the weight they bear. Lucy's small lower back bone joint surface areas and her large arm bone shaft cross-sections are, therefore, inexplicable if she moved about without routine use of her forelimbs for support.

The angle that the lower back of *Australopithecus afarensis* makes with the pelvis does not reflect the back curvature associated with human two-legged postures (*see Homo sapiens,* Skeleton, Gait, and Posture). The hip joint orientation indicates that Lucy's thigh bone could not extend as far back as in humans. When the thigh bone was extended, the kneecap would have faced to the side and not forward. These features suggest that Lucy would not have had a striding modern human-like gait in which knee flexion and extension contribute to forward movement.

Australopithecus afarensis ankle and foot bones (A.L. 288-1as, A.L. 333-8, A.L. 333-28, A.L. 333-36, A.L. 333-47, A.L. 333-54, A.L. 333-54, A.L. 333-78, and A.L. 333-79) reflect a mobile foot that lacked the fixed arches needed to support the body in two-legged postures. The big toe was not held fully parallel to the other toes, presenting a degree of divergence comparable to that of some living gorillas.

On the whole, the evidence indicates Lucy could climb trees to feed, but preferred to move on the ground. Evidence from the Dikika infant skeleton (DIK-1-1) indicates that *Australopithecus afarensis* had chimpanzee-like semicircular canals (an organ of balance found in the inner ear) and a gorilla-like shoulder blade, indicating it would have

Lucy, a female *Australopithecus afarensis*, searches desperately through the savannah for her missing three year old daughter.

used its forelimbs for weight support as do gorillas. *Australopithecus afarensis* ground movement would have included a two-legged gait, especially when feeding. Four-legged movements, however, were probably utilized when moving quickly or in heavy undergrowth.

Fossil Sites and Possible Range

Australopithecus afarensis is best known from the Sidi Hakoma, Denen Dora, and lower Kada Hadar Members of the Hadar Formation, at Hadar on the left bank of the Awash River, 310 km north east of Addis Ababa, Ethiopia. These deposits are composed of stream and lake sediments that formed during intermittent downstream blockage of the prehistoric Awash River. Volcanic activity has interlayered them with ash. Subsequently, the Ounda Hadar, Kada Hadar, and Sidi Hakoma tributaries of the Awash River have cut through the sediments, exposing them. *A. afarensis* is also known from Maka,

Cooling down on a hot summer's day, a male *Australopithecus afarensis* temporarily relaxes in the shallows of an Ethiopian lake.

75 km south of Hadar on the right bank of the Awash River, and from the Laetoli Beds upper unit in Laetoli, 10 km north of Lake Eyasi in north-central Tanzania. The Laetoli deposits are composed mostly (95 percent) of wind lain ashes and ash falls from eruptions of a nearby volcano. Geologists have divided the upper unit based on eight marker tuffs numbered 1 to 8, from deepest to most superficial; the deposits are exposed by the headwaters of the Garusi River, which cut through them. Only three (LH 2, LH 3, and LH 6) out of twenty-nine *A. afarensis* fossils known from Laetoli were actually found within the sediments; the rest are surface finds. All, however, are believed to come from the upper unit between marker tuffs 5 and 8, with the exception of two found below marker tuff 3 (LH 10 and LH 11 3), one found below marker tuff 4 (LH 12), and one found 9 m above marker tuff 8 (LH 15).

Australopithecus afarensis is also reputedly found in: 1) deposits contemporaneous with and below the Sidi Hakoma Member at Dikika, 4 km south and on the other side of the Awash River from Hadar; 2) the Nachukui Formation in west Lake Turkana, south of the Turkwel River, Kenya (*see Paranthropus aethiopicus*); and 3) sediments from Mt. Galili in the Afar depression, 82 km south of Aramis (*see Ardipithecus ramidus*), Ethiopia. None of these fossil remains are complete enough or have been sufficiently analyzed to assign to *A. afarensis* with any certainty, although they are from more or less the same time interval. Fossils from Fejej, Ethiopia, that have been assigned to *A. afarensis* are discussed with *Australopithecus anamensis*.

Age

Volcanic deposits at the base of the Sidi Hakoma, Denen Dora, and Kadar Hadar Members yield radioisotope dates of 3.4 million, 3.22 million, and 3.18 million years, respectively. The Kadar Hadar Member shows a disconformity (a period of erosion, in which there was no deposition) between ash layers dated at 2.92 million and 2.33 million years. All Hadar *Australopithecus afarensis* fossils are found below the older of these two ash layers and above the base of the Sidi Hakoma Member. They are thus bracketed in a time interval between 2.92 million and 3.4 million years ago. One of the Dikika *A. afarensis* fossils (DIK-2-1) comes from sediments 20 m below the Sidi Hakoma Member, and is thus older than 3.4 million years. The other (the DIK-1-1 child) comes from sediments contemporaneous with the Sidi Hakoma Member and is 3.22 million to 3.4 million years old.

A layer of ash (or tuff) at the boundary between the Laetoli upper and lower units produced a date of 3.76 million years, and the marker 8 tuff at the top of the Laetoli bed upper unit produced a date of 3.46 million years. All Laetoli fossils would thus seem to be bracketed within a time interval of 3.46 million to 3.76 million years. Ash just below marker 7 tuff, however, also yielded dates of approximately 3.76 million years. This indicates that many of the Laetoli fossils, including the marker tuff 7 footprints (*see* below, Laetoli Foot Prints), may be closer to the older dates. Maka fossils are found between an underlying ash layer dated at 3.75 million years, and an overlying ash layer that has resisted dating. Similarities between volcanic glass samples in this undated, overlying layer, and others found in the Sidi Hakoma and Tulu Bor tuffs at Hadar and Lake Turkana, were interpreted to mean all these layers were part of the same volcanic event and thus had the same age, 3.4 million years. The Maka fossils, therefore, seem to be bracketed between 3.4 million and 3.75 million years, although given their depth in the deposit they are probably closer to the younger of the two ages. Turkwel *Australopithecus afarensis* comes from yet unnamed sediments in the Nachukui Formation correlated to ash layers that yielded radioisotope dates between 3.2 million and 3.58 million years. Unfortunately, the correlations are far from certain and these dates are at best provisional. Fossils from Mt. Galili in Ethiopia come from deposits between ash layers with radioisotopic dates of 5 million and 1.4 million years.

Tools

Australopithecus afarensis remains predate tools by approximately half a million years.

Laetoli Foot Prints

The lower part of marker 7 tuff at Laetoli preserves the footprints of a number of different animals, including ones that were made by individuals walking on two legs. There are a total of three of these two-legged trails, a single one (trail A) and two side by side (trail G1 and G2/3), both at separate localities. It has been speculated that the G2/3 trail contains a smaller set of prints within the larger set, potentially indicating three sets of contemporary prints at site G. The trails at both localities show no signs of associated upper-limb contact with the ground, and the site G trails are considered by most paleoanthropologists to have been made by bipedal hominids. One of the site G trails preserves enough detail to allow comparisons to modern humans. These comparisons show the steps are much shorter and the feet have more pronounced toe-out than in modern humans. Assuming modern human relationships between stature and foot length and between step length and walking speed, the Laetoli trail-maker walked slower than is normal for humans. There is no reason, however, to assume that this individual would have conformed to modern human step length to walking speed relationships. The trails could just as easily reflect an individual that had proportionately longer feet and shorter lower limbs than humans and walked with much quicker steps at faster speeds than the human model predicts. Unfortunately, the foot prints do not preserve enough detail (neither toe to foot length relationships nor plantar pad and ridge patterns) to conclusively decide if they were made by a hominid. Because chimpanzees and orangutans may walk on two legs and their two-legged trails (commonly seen along river banks) resemble those made by humans, the Site G trails could have been made by *Australopithecus afarensis* or any other member of the human/great ape lineage. While the presence of *A. afarensis* fossils in the Laetoli deposits does make it a more likely candidate, the trails provide no conclusive evidence as to whether their maker's two-legged gait was as common as it is in humans or as rare as it is in great apes.

Animals and Habitats

About forty percent of the animals found with *Australopithecus afarensis* at Hadar represent genera that are now extinct. These include a leaf-eating monkey (*Rhinocolobus*), an ancient baboon (*Parapapio*), a giant otter (*Enhydriodon*), a hunting hyena (*Chasmoporthetes*), a prehistoric laughing hyena (*Percrocuta*), a false saber-tooth (*Dinofelis*), the dirk-tooth and scimitar cats (*Megantereon* and *Homotherium*), an extinct giant porcupine (*Xenohystrix*), an extinct north African/Asian mouse (*Saidomys*), an unidentified species of wooly mammoth (*Mammuthus*), an elephant with lower jaw tusks (*Deinotherium*), a three-toed horse (*Hipparion*), three bush pigs (*Kolpochoerus*, *Nyanzachoerus*, and *Notochoerus*), a short-neck giraffe (*Sivatherium*), two buffalos (*Ugandax* and *Pelorovis*), a roan relative (*Praedamalis*), a hartebeest relative (*Damalops*), and a musk ox relative (*Makapania*). Animals belonging to living genera include a leaf-eating monkey, an African mole-rat, an Indian bush rat, two species of Indian soft furred rats, a multi-mammate mouse, a porcupine, a black-backed jackal, a striped polecat, an African striped weasel, a slender tailed mongoose, an African mongoose, a civet, a laughing hyena, a brown or striped hyena, a cheetah, a serval, a lion, two Indian elephant species, two white rhinoceros species, a black rhinoceros, two giraffes, two bushbuck species, three kob species, an impala, a dik-dik, a steenbok, and a gazelle. A leaf-eating monkey, a pygmy antelope, and two gnu/wildebeest species resisted generic identification, and it is not known whether they represent extinct or living genera.

As is characteristic of many of the Ethiopian early fossil hominid sites, some of the animals (i.e., the rats and civet) have Asian affinities. There are few strictly forest-dwelling animals. The wide variety of carnivores is typical of grassland and woodland mosaics, with heavier brush and closed woodlands along watercourses and in mountain valleys.

Animals not present in Hadar that are present in Laetoli reflect both the more arid conditions at Laetoli and the probable presence of forests in highlands above 2400 m. These include a leaf-eating monkey (*Paracolobus*), a rabbit (*Serengetilagus*), a false saber-tooth (*Dinofelis*), a wolf (*Megacyon*), a prehistoric relative of horses and tapirs with claws instead of hooves (the chalicothere *Anclyotherium*), prehistoric cattle (*Brabovus*), and an extinct hartebeest (*Parmularius*). A forest duiker that has resisted classification may very well represent an extinct genus. Living mammal genera present only at Laetoli include an elephant shrew, a bush baby, a baboon, a ground squirrel, an African bush squirrel, spring hare, a naked mole-rat, a gerbil, an African climbing mouse, a fat mouse, a pouched mouse, an Acacia rat, a meerkat (or yellow mongoose), a honey badger, an oriental civet, a dwarf mongoose, a banded mongoose, a fox, a bat-eared fox, an aardvark, an African elephant, a bush pig, a camel, and a rhebok.

Analysis of fossil pollen shows that the vegetation when *Australopithecus afarensis* inhabited Laetoli was much as it is today, i.e., open *Acacia* woodlands, bush, and grasslands. Mountain forest existed in the nearby highlands. Because fossil pollen at Laetoli shows the presence of vegetation commonly found at elevations between 1500 m and 1800 m, it confirms that this elevation has remained more or less the same since the time of deposition.

Climate

Located 74 km north of Aramis and at only 100 m lower elevation (625 m vs. 520 m), the climate at Hadar is similar to that at Aramis (*see Ardipithecus*). Only 20 km southeast of Olduvai Gorge, the Laetoli and Olduvai climates are in most respects the same. Because elevations at Laetoli may be 200 m to 360 m higher, temperatures on average are 2.3° C lower than at Olduvai. Despite that, Laetoli and Hadar receive similar annual rainfall; a mean annual evaporation rate (due to a higher elevation) at Laetoli that is four times the mean annual rainfall (2000 mm vs. 500 mm) creates much more arid conditions than at Hadar. There is no evidence that the climate when *Australopithecus afarensis* lived at Laetoli would have been any different than it is today.

Classification

Despite being represented by relatively complete remains, *Australopithecus afarensis* is plagued by classification issues. These can be summarized as follows: Do the *A. afarensis* fossils belong to more than one species; to *Australopithecus*; or to a human lineage exclusive of African apes? The facts that Laetoli fossils span a time interval different from those from Hadar and that the two sites are separated by more than fourteen degrees latitude decreases the likelihood that the two belonged to the same species. Very few living African primate species have a range that extends from south of the equator to north of ten degrees latitude. Because Laetoli *A. afarensis* fossils consist mainly of isolated teeth and some upper and lower jaw bones, and such fragmentary specimens can not always be used to tell human and great apes species apart (*see Homo pekinensis* for confusion of human and orangutan teeth), the question as to whether the Laetoli and Hadar fossils belong to the same species is set to be a subject for ongoing debate. A debate that will be further exacerbated by the stong likelihood that the Hadar *A. afarensis* fossils encompass two species, as seems probable given the 500,000 year interval the Hadar deposits span and the marked differences (not solely limited to size) these fossils exhibit, e.g., the A.L. 288-1 partial skeleton vs. the A.L. 333 skeletal remains. It may prove difficult, therefore, to decide which fossils from Hadar, if any, belong to the same species as the incomplete lower jaw from Laetoli that was designated as the *A. afarensis* holotype specimen.

The belief that *Australopithecus africanus* shares a relationship with *Paranthropus* that is not shared by *Australopithecus afarensis* has prompted some to refer *A. afarensis* to another genus, *Praeanthropus afarensis*, using the genus name given to the Laetoli fossils by Edwin Hennig (*see* below, Historical Notes). This is based on the notion that, if *A. africanus* is more closely related to *Paranthropus robustus*, both of which have a different generic name, then surely *A. afarensis*—which is not as closely related to either—should also have a different generic name.

Not all would agree, however, that *Australopithecus africanus* is closer to *Paranthropus* than is *Australopithecus afarensis*. Unfortunately, most comparisons of *A. africanus*, *A. afarensis*, and *Paranthropus* with classification in mind have been confined to the skull and teeth. Other skeletal characteristics have been all but ignored. *Australopithecus afarensis* does not exhibit features by which it can be conclusively determined whether it should be placed in the genus *Australopithecus* or whether it is even an exclusive member of the human lineage.

Australopithecus afarensis shares front tooth and palate characteristics and outer brain blood circulation with orangutans exclusive of African apes and humans. At the moment it is not certain whether these features were inherited from a common ancestor with orangutans, or whether they developed in orangutans and *A. afarensis* independently. Because some mammal species at Hadar show an Asian connection, and orangutan-like apes existed in southern Asia and Arabia 9 million to 14 million years ago, an Asian affinity, exclusive of African apes and humans, for *A. afarensis* is another possibility to consider.

Historical Notes

Louis Leakey is credited with finding the first remains of *Australopithecus afarensis* at Laetoli, a left lower canine (BMNH M.18773), in 1935. This find remained unidentified in the collections of the British Museum of Natural History until it was recognized as *A. afarensis* in 1979. In 1939 Ludwig Kohl-Larsen's expedition to Laetoli uncovered an isolated molar and an upper jaw with two molars. In a 1948 issue of *Wissenschaftliche Rundschau*, Edwin Hennig referred these remains to the genus *Praeanthropus*. Muzaffer Senyurek followed Hennig in a 1955 issue of *Belleten*, and also resuscitated a suggestion by Hans Weinert that the species name should be *Praeanthropus africanus*. In October 1973, Donald C. Johanson found the first remains of *A. afarensis* at Hadar: two upper thigh bones (AL 128-1 and A.L. 129-1a), fragments of a lower thigh bone (A.L. 129-1b), and an upper shin bone (AL 129-1c), preserving the knee joint. Together with a skull fragment, these fossils were described by Maurice Taieb and colleagues the following year in a 1974 issue of *Comptes Rendus de l'Académie des Sciences*. That same year, Mary Leakey began excavations at Laetoli that would last until 1979. In a March 1976 issue of *Nature*, Johanson and Taeib reported the A.L. 288-1 skeleton, and interpreted similarities between the Hadar fossils and 1) *Australopithecus* (Sts 14 and A.L. 288-1); 2) *Paranthropus* (A.L. 211-1 vs. SK 87, SK 92, and OH 20); and 3) *Homo* (A.L. 200-1a, A.L. 266-1, and A.L. 277-1 vs. Sanigran IV, OH 7, KNM-ER 1590, and KNM-ER 1802) as evidence that all three genera were contemporaneous 3 million years ago in the Hadar deposits. In 1977 and 1980 issues of the *American Journal of Physical Anthropology*, Tim White published descriptions of the Laetoli fossils recovered in 1974-1975 and in 1976-1979. In a 1978 issue of *Kirtlandia*, Johanson, Yves Coppens, and White designated the LH 4 adult lower jaw from Laetoli as the holotype for their new species *Australopithecus afarensis*. Three years later, in the *South African Journal of Science*, White, Johanson, and William Kimbel defended their claim that *A. afarensis* was the earliest human ancestor by arguing that *Australopithecus africanus* was an evolutionary side branch of the human lineage. The

It's a brand new day for Lucy's child 3.4 million years ago.

species name *afarensis* refers to the Afar region of Ethiopia where most of the fossils of this species were found.

The April 1982 issue of the *American Journal of Physical Anthropology* (volume 57, #4) was dedicated to describing the ensemble of Hadar *Australopithecus afarensis* fossils found between 1974 and 1977. In a 1983 issue of *Folia Primatologica*, Peter Schmid noted that *A. afarensis* had a funnel-shaped thorax, broad hips, and a generalized body plan showing neither human nor living great ape specializations. In a 1983 *American Journal of Physical Anthropology* issue, Jack Stern and Randall Susman challenged Johanson and colleagues' earlier claims that *A. afarensis* was routinely two-legged and ground-moving. They emphasized its tree climbing ability by pointing to similarities with chimpanzees. Many of these similarities had already been alluded to by Christine Tardieu and Bridget Senut in their Doctoral theses. In a 1985 meeting of the American Association of Physical Anthropologists at Knoxville, Tennessee, Esteban Sarmiento, noting that chimpanzees were partly terrestrial and partly arboreal, suggested that the *A. afarensis* chimpanzee-like characteristics could just as well be associated to four-legged movements. This was followed by a series of papers published in the American Museum of Natural History *Novitates* showing African apes and humans shared a four-legged ground moving phase in their evolutionary history, and A.L. 288-1 (Lucy) would not have routinely stood on two legs because she lacked those skeletal features that develop during an individuals lifetime in response to two-legged stance and movement. In a March 1994 *Nature* issue, Kimbel, Johanson, and Yoel Rak reported the A.L. 444-2 skull and new dates for the Hadar fossils, making Al 288-1 approximately three hundred thousand years older than the dates published a decade earlier. In a 2000 *Evolutionary Anthropology* issue, Stern reiterated the arboreal nature of *A. afarensis*, though in a 2005 issue of *Gait and Posture*, Owen Lovejoy presented evidence that *A. afarensis* routinely stood on two legs.

DISPATCHING A MORTAL ENEMY

Weaver birds flew by the dozens back and forth to the palm trees, stripping the fiber from the palm fronds to build their ball-like nests. The noisy chatter that accompanied their labor drowned out all other sounds. A troop of apemen sat below the palms. They fed silently on the ginger bread fruit that had fallen from the palms. They ate both the pulp and kernel, leaving behind only the husk. Exhausting the fruit supply on the ground, some of the apemen climbed up to drop fruit down to those below. Other apemen fed on the terminal buds of young palm trees that sprouted from suckers at the base of the big palms. By mid-morning the apemen had finished off the bread fruit that was ripe and headed back through the grassland to the lake edge. As they neared the lake, they could hear the trumpeting sounds of elephants bathing along the shore. Following a warthog trail through shoulder high elephant grass that grew along the lake floodplain, the apemen reached the lakeshore a good distance away from the elephants. Here the shoreline was covered in woodland, and the shaded ground nurtured succulent herbs. A large-leafed herb with cherry-like fruit that the apemen relished grew in abundance. They ate the plant's fruit, leaves, and roots and by mid-afternoon left to feed elsewhere.

The apemen ambled up a shaded streambed, which was nearly dry. There was no running water, although the deeper stream sections still harbored puddles. Walking along the sandy bed, the apemen stood on two legs to traverse puddles and other areas of the streambed that were soft and muddy. Walking around onto the stream's grassy bank, to avoid one of the larger puddles, one female apeman shrieked loudly. Agitated at what she saw, she jumped up in the air, changed directions, and ran back on all fours to the stream bed. Picking up a large stick, she reared up on two legs and strode back to the source of her fright. Once there, she furiously beat at a sapling with the stick, all the while grunting and groaning. Other apemen who saw and heard the commotion ran to assist, and repeatedly beat at the ground. Standing on two legs, one of the large males lifted a boulder above his head with both hands and threw it down to the ground with all his might. The boulder crushed the head of a snake, which now lay dead with its body limp and kinked from the beating. The apemen patiently prodded the snake's body with sticks, testing for movement. A twitch of the snake's tail elicited shrieks from some and prompted another round of beatings. Finally, certain it was dead, one of the apemen grabbed it by the body and threw it into some shrubs away from the stream.

With the evening quickly approaching, the apemen constructed nests from shrubs and herbs growing on the shaded woodland floor. Some of the youngsters made their nests in small trees above where the adults slept.

Sitting up in her nest with the stick still in her hand, the apemen stared suspiciously at a small sapling within arm's reach. Cocking the stick with her right and bending the sapling with her left, she carefully inspected the foliage for snakes. When none were found, she put the stick down and stripped the sapling's bark. Getting at the underlying pith, she enjoyed a final bite before turning in for the night.

PARANTHROPUS AETHIOPICUS

Skull, Teeth, and Diet

Fossils classified as *Paranthropus aethiopicus* are mostly isolated teeth that provide limited information. The Black Skull (KNM-WT 17000) is the best preserved of the known fossils and thus provides most of what is known about the appearance and adaptation of this species. As in other *Paranthropus* species, its canines are small, and the cheek teeth are large with flat wear. The Black Skull cheek tooth rows, however, are long and set parallel to each other, and there is a wide space between right and left canines, suggesting wide front teeth. The Black Skull's features strongly contrast with the forward converging cheek tooth rows and narrow front teeth of the *P. aethiopicus* holotype specimen (Omo 18-1967-18), a toothless lower jaw bone, and serve to distinguish the two. A wide front tooth row also distinguishes the Black Skull from all other *Paranthropus* species.

The Black Skull cheek bones are very wide from side to side and top to bottom, attaching low on the upper jaw at the level of the first premolar. They thus have a forward set relative to the upper jaw. This gives the face a flat-dished appearance with only a moderate snout. Given that the tooth row is long and protrudes forward, the plane of the face is inclined upwards. The face is large relative to the braincase, and there is a marked narrowing of the skull behind the eye sockets. The upper midline of the braincase has a large crest for the attachment of chewing muscles. In living apes a crest of corresponding size is a feature seen only in males. In keeping with its wide face, its skull base is also wide and resembles that of other *Paranthropus* species. The braincase, however, is narrow and has an estimated brain volume of 410 cc, which is below the average of other *Paranthropus* species. A damaged and incomplete juvenile skull (L338-y), which has also been referred to *P. aethiopicus*, has a slightly higher estimated volume (427 cc).

There are no reported studies on the diet of *Paranthropus aethiopicus*. Its massive cheek teeth suggest a bulk diet low in nutrition and high in volume, consisting largely of greenery and roughage. Considering their suggested size, the front teeth were probably important in processing food, stripping the bark and stems to get at the succulent pith inside. Given a wet grassland habitat (as suggested by the lake and river deposits in which it is found), roots and tubers easily dug out by hand from wet soil were probably an important part of its diet. Fruits would have been eaten whenever available.

Skeleton, Gait, and Posture

There are no skeletal bones directly attributable to *Paranthropus aethiopicus*. There are, however, some unidentified fossil bones of corresponding body size from the same deposits and time interval in which the *P. aethiopicus* skulls and teeth were found. These fossil bones may provide insight into how this early hominid moved.

The upper half of an arm bone (Omo 119-1973-2710) has a shoulder joint surface, which is small relative to the shaft's thickness. The bone's relative joint size, large muscle attachment areas, and the leverage these attachment areas afford all suggest the forelimb was used for weight support and

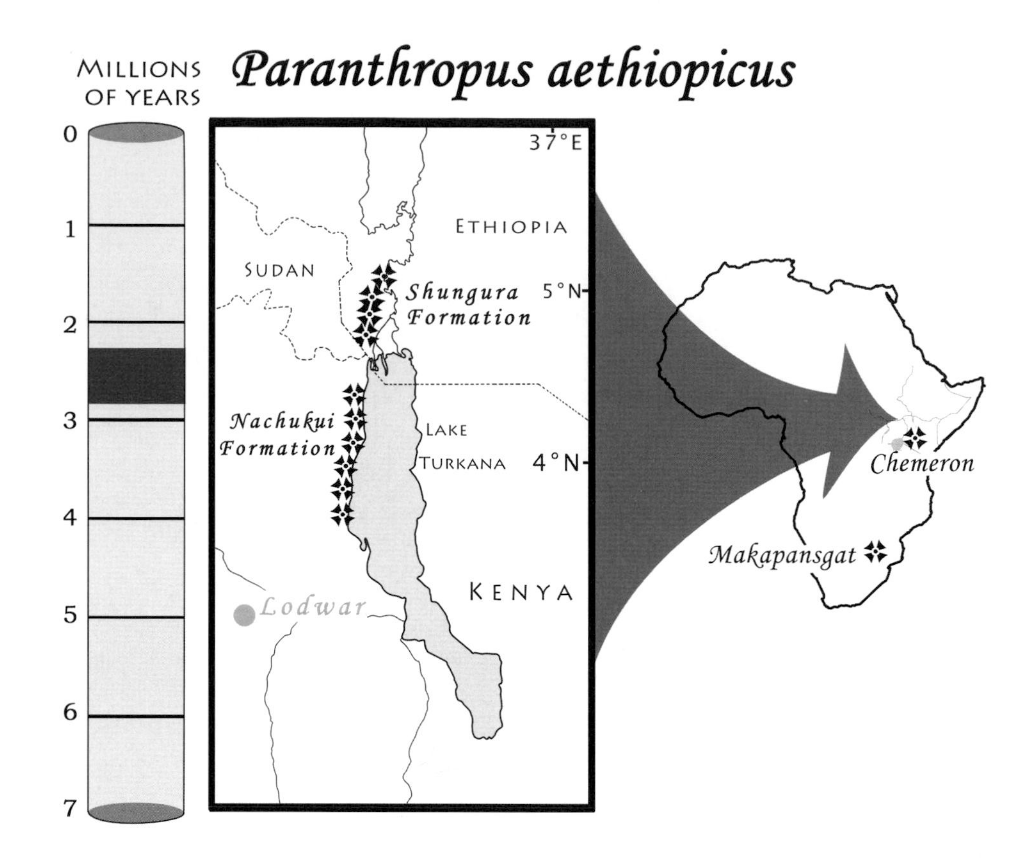

propulsion. Measurements on a damaged but nearly complete forearm bone (L40-19) show it to be closer to living chimpanzees than to any other great ape or human. Its shaft curvature, joint surfaces, and length (just below the upper length range for chimpanzees) suggest a multi-purpose forelimb employed for moving both on the ground and within trees. A complete heel bone (Omo 33-1974-896) shows the foot lacked the fixed human-like arches and was much more flexible than human feet. All the fossils are consistent with an early hominid that spent most of its time on the ground, but retained tree-climbing abilities. Supposing that at least some of these bones belong to *P. aethiopicus*, it is likely that this species would not have moved habitually on two-legs as do humans.

Fossil Sites and Possible Range

Paranthropus aethiopicus fossils are found in lake and river deposits that compose the Nachukui and Shungura Formations on the west side of Lake Turkana, Kenya, and the Omo River basin, Ethiopia. With its source situated in the highlands of Ethiopia, the Omo River flows southward towards Kenya, blindly discharging into Lake Turkana at the western border of the two countries. Losing speed as it nears the lake, the Omo River deposits its sediment. Over millions of years the sediment from this river and other smaller rivers that discharged in the basin has accumulated and resulted in the fossil bearing formations. Intermittent eruptions from nearby volcanoes, throughout the time interval the deposits were accumulated, layered the formations with volcanic ash. The heaviest ash layers divide each formation into different members. *P. aethiopicus* is recognized from Members C-F of the Shungura Formation and from the Lokalalei Member and Tuff of the Nachukui Formation, Kenya. Because a large part of the *P. aethiopicus* remains are from river deposits in the Shungura Formation, the fossils are damaged by water action and most fossils are very fragmentary and consist of isolated teeth. A skull fragment, preserving the jaw joint and ear canal (KNM-BC1) from the Chemeron Formation in the Tugen Hills, Kenya (*see Orrorin tugenensis*), considered by some to be *Homo*, probably represents *P. aethiopicus*, suggesting this species had a wide distribution.

Age

Radioactive and paleomagnetic dating of the ash layers provide absolute dates for when the deposits were formed. *Paranthropus aethiopicus* is recognized from a time span between 2.3 million and 2.8 million years ago.

Tools

There are no tools directly associated with *Paranthropus aethiopicus*. Stone tools, however, have been found in deposits spanning the time intervals *P. aethiopicus* existed (i.e., in Member F and supposedly in Member E of the Shungura Formation, possibly dating back to 2.4 million years, and in Lokalalei Member of the Nachukui Formation, Kenya, dating back to 2.34 million years).

At one Lokalalei tool locality are stone cores that show evidence of repeated flaking. These suggest the toolmakers attempted to maximize the number of flakes that could be produced per stone unit. Because these tools are not associated with any other hominid species, it is possible that they were made and/or used by *P. aethiopicus*. There is some uncertainty, however, as to whether the deposit in which the older tools are found actually correspond to Member E, and whether or not some of the supposed Member E tools were swept in by water from overlying younger deposits. If *P. aethiopicus* is responsible for the tools and these do not appear until Member F times, it was only in the last 40,000 years of this species' lifespan that it used them.

Animals and Habitats

Most animals found in the *Paranthropus aethiopicus* deposits are modern in appearance. There are shrews, three different bat genera, a bush baby, two species of gelada baboons, a hare, a ground squirrel, a bush squirrel, a gerbil, nine different rat and mice genera, a jerboa, a cane rat, a naked mole-rat, two porcupine species, a slender

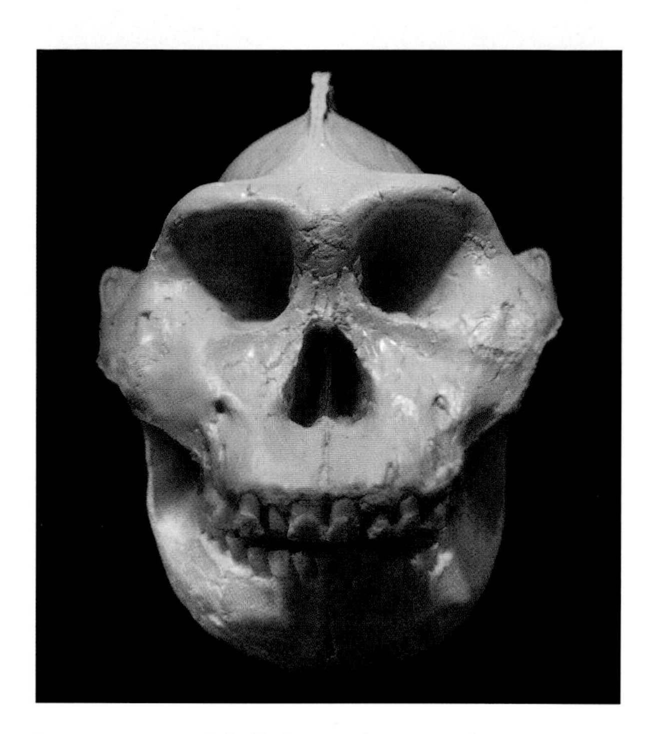

Reconstruction of skull of *Paranthropus aethiopicus* based on the Black Skull, using a modified Pening 1 mandible.

and dwarf mongoose, a striped hyena, a leopard, a robust species of African lion, a caracal, an African elephant, a close relative of the Indian elephant, an aardvark, both the white and black rhinoceros, a camel, a giraffe, an extinct species of cape buffalo, three species of kudu/bushbuck, two species of kob/waterbuck, and an impala. Extinct genera include a leaf-eating monkey (*Paracolobus*), a prehistoric baboon (*Parapapio*), two saber-tooth cats (the dirk-tooth *Megantereon* and the scimitar cat *Homotherium*), a large cat more closely related to house cats than to lions (*Dinofelis*), a hyrax (*Gigantohyrax*), an elephant with lower jaw tusks (*Deintotherium*), a plant-eating cousin of horses and tapirs with large claws instead of hooves (*Anclyotherium*, a chalicothere), two pigs (*Kolpochoerus* and *Metridiochoerus*), a short-neck giraffe (*Sivatherium*), two species of a waterbuck a kob relative *(Menelikia)*, and a musk ox, *(Pelorovis)*. A horse, a roan/sable, a gnu/wildebeest, a bison/cattle, and a goat could not be identified and may also represent different genera. There also appears to be another ape or human in the *P. aethiopicus* deposits, which is likely to represent either *Australopithecus garhi* or *Homo*.

Despite a wide variety of animals found in the *Paranthropus aethiopicus* deposits, reflecting a wide range of habitats, absence of those mammals that today inhabit forests throughout Africa, e.g., tree hyraxes, pouched rats, guenon monkeys, duikers, tree pangolins, and brush-tailed porcupines, makes it unlikely that *P. aethiopicus* lived in forests. Because the Black Skull has minor damage and is relatively complete, it is likely to have been deposited close to where it died. The high percentage of waterbuck and reedbuck from the Lokalalei deposit and the presence of two species of *Theropithecus* at both Lokalalei and Shungura E suggest *P. aethiopicus* inhabited flooded grasslands.

Climate

There has been much speculation as to climatic changes during the time interval when *Paranthropus aethiopicus* lived based on differences in fossil animals found in the deposits. With highlands to the north, southeast, and west of Lake Turkana and the Omo Basin already formed 3 million years ago, the weather during the time *P. aethiopicus* lived must have been very similar to what it is today. Changes in river and stream courses and intermittent formation of lakes in the Omo Basin area, as opposed to changes in rainfall over time, best explain why fossils from animals associated with different habitats and varying rainfall appear at different times in the formations.

Classification

The holotype specimen lower jaw of *Paranthropus aethiopicus* (Omo 18-1967-18) is battered and nondescript, and lacks all the tooth crowns. In this condition, it has proved problematic for *P. aethiopicus* classification. Without apparent features to compare to this jaw, the Black Skull was assigned to *P. aethiopicus* based on geographic proximity and similarities in body size and geologic age. The assumption made was that only one hominid species with a given body size existed at any one time in any one area, and individuals with markedly different body sizes cannot belong to the same species. Given that 1) gorilla males may be two to three times as large as gorilla females, 2) gorillas, chimpanzees, and humans have overlapping body sizes, and 3) all three coexist in the western rift forests less than eight hundred kilometers away from Lake Turkana, basing classifications on such criteria is questionable.

In turn, the Black Skull and the holotype specimen jaw were assigned to *Paranthropus* based on similarities between the Black Skull and the more recent skulls of *Paranthropus robustus* and *Paranthropus boisei*. This opened the way for isolated teeth with *Paranthropus boisei*- and *Paranthropus robustus*-like molar features, from the same time interval and formation as the holotype specimen jaw, to be classified as *P. aethiopicus*. Presenting only a partial worn premolar crown, however, there is no reason to believe the Black Skull has *P. boisei*– or *P. robustus*–like molars, or that its teeth would be anything like those of the toothless holotype jaw. In fact, differences in tooth row set and front tooth width suggest the holotype jaw and the Black Skull are different species. It remains to be seen whether the isolated teeth grouped in the same species with the holotype jaw actually belong to this species, to the Black Skull, or to some other species.

Considering their similar appearance and geologic age, the skull fragment from Chemeron in the Tugen Hills west of Lake Baringo may prove to be the same species as the Black Skull. The same applies to a juvenile lower jaw from Makapansgat (MLD 2), which shares a large number of features with and has approximately the same geologic age as the lower jaw (KNM-WT 16005) from the

Amused by the playfulness of his children, an adult *Paranthropus aethiopicus* watches protectively and lovingly.

same deposits as the Black Skull. For now, there is tantalizingly little evidence available for deciding which fossils should be included in the same species as the Black Skull.

Historical Notes

In August 1967, Canille Arambourg and Yves Coppens gave the name *Paraustralopithecus aethiopicus* to a damaged and toothless jaw bone (Omo 18-1967-18) found by Rene Houin in Member C of the Shungura Formation in July of the same year. Considering it was the first fossil hominid found in Ethiopia, the authors chose the species name "*aethiopicus*" to honor the country. In August 1986, Alan Walker, Richard Leakey, Jack Harris, and Frank Brown described the Black Skull and a lower jaw (KNM-WT 16005), both found in deposits west of Lake Turkana in 1985, as *Australopithecus boisei*. Subsequently, in 1987 Francis Clark Howell, Paul Haeserts, and Jean de Heizelin, heeding a suggestion made by Walker and colleagues, referred the Black Skull and the lower jaw to *Paranthropus aethiopicus*. They also included in this species assorted teeth and jaws from Members E and F of the Shungura Formation and the partial juvenile skull (L 338-y), which Rak and Howell described in 1978 as *Australopithecus boisei*. In 1996, Gen Suwa, Tim White, and Howell referred all Shungura Members C-F molars that were *A. boisei*-like to *Australopithecus aethiopicus*.

THE FIRST TOOL-USING SCAVENGER?

A tremor and then an explosion rocked the land. Shortly thereafter the mid-day sun was blocked by a black cloud streaming out from the mouth of a volcano. The cloud grew larger and larger, making night out of day. A strong wind blew a mixture of water and ash, which rained down on the land until the early hours of the following day. Throughout the storm, animals stood motionless, and the birds were silenced. A group of apemen took cover in a grove of forest fever trees growing in a flooded glade adjacent to a river. When the ash finally stopped falling, the red glow of two active volcanoes could be seen on either side of the horizon.

The morning sky was overcast. The thick blanket of ash that had descended the day before had dusted both animals and vegetation in a light grey hue so that the still motionless animals looked like garden statues. The apemen descended from their nests, shaking loose some of the ash covering their body. Dusting off a large white flower before putting it in his mouth, one of the apemen scouted the land and headed in the direction of the stream. Reaching the stream edge, he sat on the ash covered ground, pulling cat tails out by their roots. Stripping the leaf blades with his hands, he used his front teeth to harvest the pith. As he ate, he paid little notice to the ash-covered body of a giant hartebeest floating in the center of the stream. By late morning he tired of eating pith and moved on to join the rest of his group. Walking upstream, he occasionally stopped to pull out and nibble on plants that grew along the stream edge.

One week had passed since the explosion, and the land still lay blanketed in ash. The ash had killed many of the smaller plants, and very little food was now available. The heavy rains that finally came washed away the ash and caused the rivers and streams to swell. The swollen waterways washed away the bodies of animals that had died from the vapor and heavy ash fall in the immediate surroundings of the volcanic eruption.

A bush pig rooted among the bodies of a three-toed horse and a hartebeest that had collected in the outside bend of a river. Standing on two legs and wading out into the water, a female apeman appeared from behind a reed bed and dragged the horse's carcass belly up onto the shore. Hitting two stones together, the apeman produced a sharp flake, which she used to disembowel the horse. The strong smell of the guts repulsed the apeman, and she left these to rot on the shore. The meat, however, from the trunk and limb muscles was aged just right. Wielding a stone tool, the apeman tore into the flesh with relish, satisfying her hunger. In no time, two dozen or more apemen descended on the horse's corpse. Brandishing sharp stones, they cut at the corpse and pulled the horse apart at the joints. Hooting, they carried away a piece of meat and bone to a secluded area to peaceably enjoy their meal. A mother and her child ran off with one of the legs. The mother quickly stripped off the meat and ate it. Banging the cannon bone with a hammer stone, the mother broke the bone in two to get at the marrow inside. She gave this to her child and proceeded to break open the other long bone and drain it of marrow. After the meal, the apemen went their separate ways, dispersing into smaller groups. By late afternoon three bush pigs where feeding on the remains of the carcass. The apemen were nowhere in sight.

AUSTRALOPITHECUS GARHI

Skull, Teeth, and Diet

An upper jaw with teeth and some associated skull fragments, a lower jaw with teeth, two toothless lower jaw fragments, and three small skull fragments are all that has been described for *Australopithecus garhi*. The teeth are similar in size and appearance to those of *Australopithecus africanus* from Sterkfontein. As in *A. africanus*, the molars are large, with a generalized great ape cusp pattern and poor development of molar crests associated with leaf-eating. Poor crest development, however, does not indicate *A. garhi* did not eat leaves, but reflects the incompatibility of molar crests with chewing hard foods (crests are fragile and break under increased pressure). By increasing the food amount processed with each jaw stroke, large molars can increase efficiency without molar specializations committed to any one food type. As in *A. africanus*, molar size and appearance suggest a varied diet low in nutritious content and processed in bulk. The canine is very stout, with a large x-section and root, but does not project appreciably past the tooth row chewing surface and shows only tip wear. There is a slight gap (diastema) between the upper canine and front teeth. Although the described A. *garhi* front teeth are damaged, they also appear to be within the *A. africanus* size range, suggesting they had a varied function in preparing food for molar grinding. Sculpting the missing portions of the *A. garhi* braincase resulted in a 450 cc estimate of brain volume.

Skeleton, Gait, and Posture

All that has been reported for *Australopithecus garhi* are two separate arm bone fragments, the lower half of a forearm bone, and an associated fragmentary skeleton, consisting of a toe bone, a shin bone fragment, and the shafts of the left thigh bone, the right arm bone, and two forearm bones. Unfortunately, none of these are associated with teeth or skull remains that can be attributed to the *A. garhi* species. Only a few of these bones preserve joint surfaces.

Despite incomplete and unassociated fossils, the *Australopithecus garhi* limb proportions were interpreted to be more human-like than those of *Australopithecus afarensis* (A.L. 288-1, Lucy). With neither the arm bone nor thigh bone shaft fragments preserving joint ends, and with unknown lengths missing from both shafts, these limb proportions are prone to considerable error. Arm bone to thigh bone mid-shaft diameter ratios of *A. garhi* differ from what is common in humans and suggest that the forelimb may have supported relatively more weight than in humans. The toe bone is similar in appearance and dimensions to those of *A. afarensis*. If the associated skeleton is *A. garhi*, the short toe bone indicates that this species moved mainly on the ground. Information as to the types of ground movement it engaged in must await recovery of more complete remains, including bones with joint surfaces. Because all great apes occasionally walk on two legs, it is fair to assume *A.*

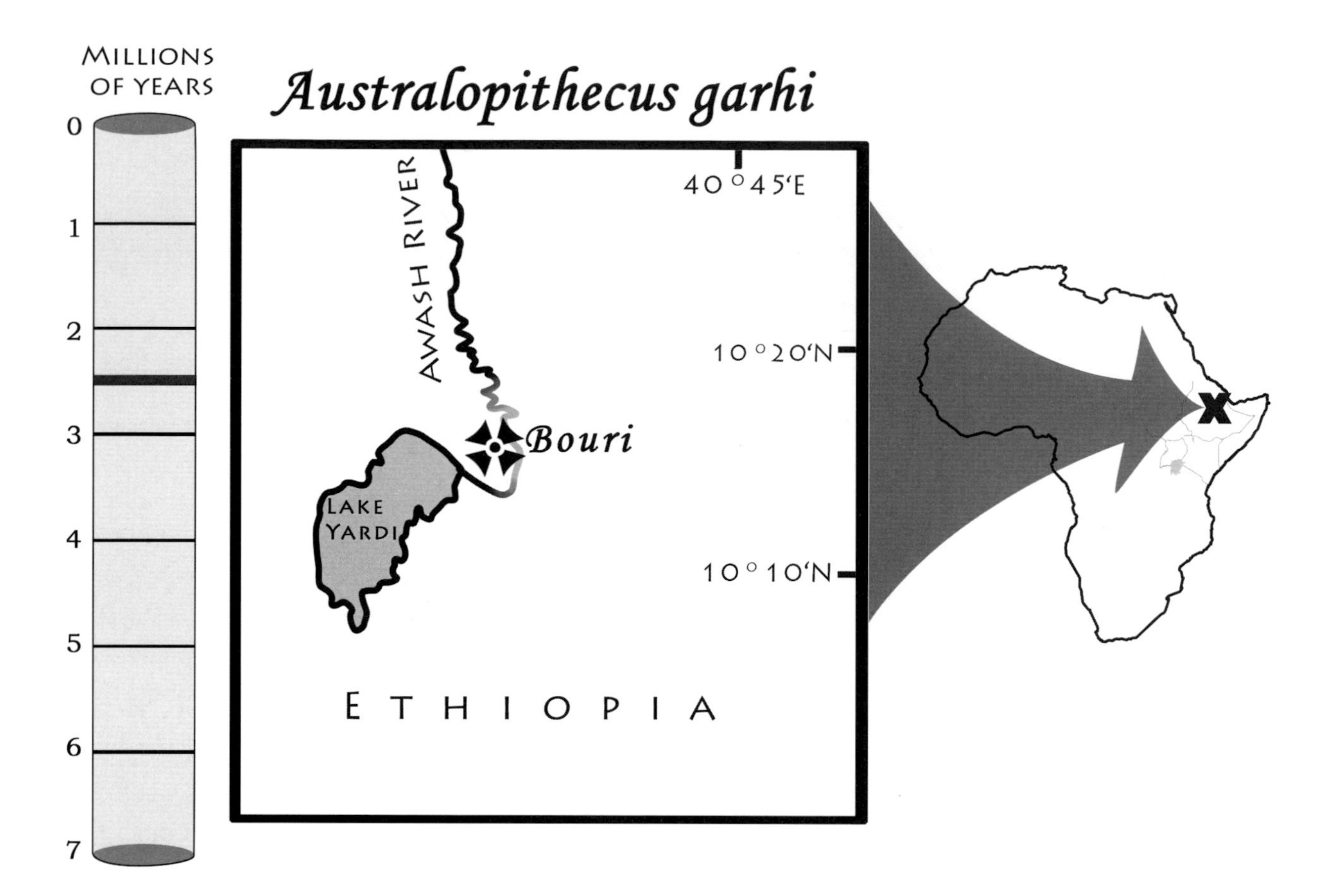

garhi did so also. Whether *A. garhi* would have spent more time on two legs than any of the living great apes is a question that cannot be answered at this time.

Fossil Sites and Possible Range

Australopithecus garhi fossils are found within the Hata Member of the Bouri Formation at three different localities: Bouri, Gamedah, and Matabaietu on either side of the Middle Awash River just north of Lake Yardi in central Ethiopia. The Hata Member represents a lake delta deposit of slow moving water made up of mudstones and sandstones. The deposits also contain ash layers from nearby volcanic eruptions that occurred at the time of deposition. Similarities between *A. garhi* teeth and the non-associated teeth of an unidentified hominid, contemporaneous with *Paranthropus aethiopicus*, suggest *A. garhi* may also come from the Members C-F of the Shungura Formation in the Omo Basin, Ethiopia. The partial juvenile skull (L338-y) from Member E, which some claim is not *Paranthropus*, may therefore represent *A. garhi*. Without more complete remains, however, attribution of fragmentary remains from the Omo Basin or elsewhere to *A. garhi* is uncertain.

Age

Radioisotope dating of one of the largest ash layers in the deposits arrived at an age of 2.5 million years. Paleomagnetic dating confirmed this result. The *Australopithecus garhi* associated skull, which was found in sediments 5 m above the ash, is somewhat younger, with an age estimated at 2.46 million years.

Differences Between Males and Females

There is no evidence as to whether the remains represent males or females, and whether there were size differences between the two sexes.

Tools

No tools were found associated with the *Australopithecus garhi* fossils. Antelope bones with cut marks and broken open by hammerstones, however, were found in the same deposits in which *A. garhi* was discovered, suggesting this hominid may have eaten bone marrow and scavenged for its food. If *A. garhi* is responsible for the tools used in processing the antelope bones, it is the earliest evidence for scavenging in the human lineage and serves to confirm Raymond Dart's claims that *Australopithecus* broke bones to get at the marrow.

Animals and Habitats

Extinct genera found in the same deposits as *Australopithecus garhi* include a scimitar cat (*Homotherium*), a prehistoric elephant with lower jaw tusks (*Deinotherium*), a three-toed horse (*Hipparion*), three pig genera (*Kolpochoerus, Metridiochoerus*, and *Notochoerus*), a short-neck giraffe (*Sivatherium*), two cousins of gnus and hartebeests (*Parmularius* and *Rabaticera*), a giant hartebeest (*Megalotragus*), a musk ox (*Pelorovis*), a relative of pygmy hippopotamus, a species of Indian elephant, two bushbuck species, and a goat (*Numidocapra*). Modern mammal genera found with *A. garhi* include a baboon, a gelada baboon, a genet, a clawless otter, a giraffe, a kob, a waterbuck, a Hunter's antelope, a cape buffalo, a kudu, a topi, a gnu, a springbuck, a gazelle, a sable or roan antelope species, and an oryx.

Compared to animals found in deposits of similar age from South Africa and from the Omo Basin, the animals associated with *Australopithecus garhi* appear to be more modern. There are no extinct monkey genera or a variety of large extinct cats. Chalicotheres, extinct cousins of horses, and tapirs are also conspicuously absent. Notably, both kudu and kob are represented by modern species that live in the Awash River area today. Overall, the animals suggest an open habitat of grass and scrub with long dry seasons. Thin woodlands and grassy glades would have existed close to lakeshores and along stream courses.

Climate

Animals found with *Australopithecus garhi* indicate a dry and mild climate not much different from what it is today at elevations between 500 m and 1000 m above sea level (*see Ardipithecus,* Climate)

Classification

With *Paranthropus aethiopicus, Australopithecus africanus*, and reputedly also *Homo* present in 2.5 million-year-old African fossil deposits, it remains to be seen whether *Australopithecus garhi* will retain its distinct status when more of its fossils are discovered. Although its describers outlined a large number of features by which *A. garhi* differs from *A. africanus*, the full range of *A. africanus* variation was not taken into account. When this range of variation is considered, there are no features that clearly distinguish *A. africanus* from *A. garhi*. Some *A. africanus* fossils (e.g., the StW 258 skull) resemble *A. garhi* more than others. Although forty degrees of latitude between *A. garhi* and *A. africanus* supports placing them in separate species, the fact that *A. africanus* may have had the wide geographic distribution characteristic of baboon species leaves open the possibility that the two could be the same species.

Historical Notes

The first *Australopithecus garhi* fossils were found in 1990 at Matabaietu and Gamedah, Ethiopia. These consisted of a small skull fragment, a toothless jaw fragment, and the lower end of an arm bone. Although it was clear from the jaw fragment that these remains did not represent a robust *Australopithecus*, the fossils were too fragmentary to identify it to species. In November 1997, Yohannes Haile-Selassie found a partial skull with an upper dentition, which Berhane Asfaw,

Tim White, Owen Lovejoy, Bruce Latimer, Scott Simpson, and Gen Suwa described, in an April 1999 issue of *Science*, as the holotype specimen of *Australopithecus garhi*. This fossil consists of an upper jaw with teeth and part of the skull vault including the brow. Because the team expected to find *Australopithecus africanus*, this discovery was unanticipated, and earned this species the name "*garhi*," which in one of the Afar languages means "surprise." All of the remains, including the holotype, are housed in the Ethiopian National Museum in Addis Ababa.

A piercing scream penetrates the silence of the ancient world as *Australopithecus garhi* helplessly watches another of his kind become a meal for a hungry lion.

THE SOUTH AFRICAN FOSSIL CAVE SITES

Australopithecus africanus • Paranthropus robustus/crassidens

Acid leaches bone-dissolving bone mineral. In a pine forest where soil acidity levels are high, a buried bone lasts little more than a decade and has no chance for fossilization. Made up of calcium carbonate, limestone preserves bone, neutralizing acid. Dissolved in water, calcium carbonate may percolate into spaces within the bone, precipitating into these spaces, and further strengthening bone. For this reason fossil bones are often associated with limestone deposits.

Billions of years ago, before animals had developed hard parts or skeletons, micro-organisms living in shallow seas accumulated magnesium and calcium carbonate, which, under saturated seawater conditions probably caused by the micro-organisms, precipitated into the dolomite formations, which cover many parts of southern Africa. With uplifting of the continent, the dolomite came to lie nearly sixteen hundred meters above sea-level, exposed to air and water action.

Slightly soluble in water, dolomite over time dissolves and disintegrates. At first rain water percolates through open cracks in the dolomite. In time, some of these cracks grow and form into caves. Water courses, finding the lowest level, make their way into these caves and further erode the dolomite. Streams and rivers enlarge caves and form outside openings. Once opened, the cave accumulates outside sediments, including animal remains. Ultimately water action will lead to the collapse of the cave roof and exposure of cave sediments on the land's surface. A limestone cave's lifespan depends in large part on the amount of water available. In very arid areas, a limestone cave usually has a very long life span and in very wet areas a very short one.

Saturated with solutes, water passing through dolomite or limestone deposits may chemically precipitate calcium carbonate in a nearly pure form called flowstone or travertine. Saturated water soaking cave sediments may cement these into an extremely hard conglomerate called breccia. Falling into caves, animal remains may be sealed within flowstone or cemented into a breccia and preserved as fossils.

Formed millions of years ago, the South African limestone caves preserve a record of the evolution of the hominids that lived in these areas during a time span of nearly three million years.

THE ANIMAL TRAP

Venus and the crescent moon hung in a clear sky. The lightning that a few hours earlier had lit up the night failed to bring rain. In the absence of rain, the lightning strikes brought fire to the high veldt. One blazed on a nearby kopje. Its warm orange glow contrasted with the cool white moonlight. In a short time the fire moved down from the grass onto the forest strip along the stream. The light from the blaze awakened an apeman sleeping in the crook of a large cabbage tree. His alarm calls aroused the group, which quickly assembled on the ground below. Descending from the tree, the apeman joined the rear of the group, moving up stream into the darkness away from the fire. Crossing a dried streambed, the group turned onto the veldt, looking for a suitable place to spend the rest of the night. Just before reaching a hilltop, they crossed a dolomite outcrop hiding the aven of a large cave. With the fire well behind them, the group disappeared to sleep in the nooks and crevices the outcrop afforded.

It had been a long time since the apeman had been there. The strong scent of cat urine had long ago made it a place to avoid. The smell was now gone, and once again it was safe. Despite the years that had passed, the place was still familiar. It brought back memories of when the apeman was still young and walked at his mother's side. Looking through the aven to see the cave's hidden treasures was always entertaining. It wasn't always possible to see inside the cave. Only when the sun reached a specific position in the sky would its rays pass through illuminating the cave floor. Once, long ago, the apeman had seen an antelope trapped deep inside. Repeatedly the antelope attempted to run up the dolomite walls and jump through the opening, but would always fall back before reaching it. The antelope's repeated bleatings elicited faint higher pitched calls from a nearby fire-thorn bush betraying the location of its fawn. The apeman remembered an older brother had grabbed the fawn by its hind limbs and whipped it against a dolomite boulder, smashing its head and silencing its calls. Biting at the antelope's skin, he twisted the limbs and separated them from the body. The brother gave the limbs to their mother, and all of them feasted on meat. It was a memorable meal since it came at a time of year when food was scarce.

The light from a cloudless blue sky flooded the veldt just before sun up, pulling the apeman away from its memories. Below, last night's fire still smoldered. Hurrying down hill to the edge of burnt land, the apeman looked for a bee's nest he had seen some days earlier. The smoke would have cleared the bees, and the honey would be there for the taking. On reaching the nest, the apeman saw a honey badger clawing at debris, its snout covered in honey. Throwing stones, he attempted to chase it away, but the animal held its ground. Standing on two legs, the apeman picked up a branch glowing hot at one end and drove it into the animal's side. The badger yelped loudly, and then with a growl turned its attention to the branch and its bearer. With a swipe of its claws, it broke off the burning end, and repeatedly lunged at the branch breaking off a piece each time the apeman would jab him. Realizing the hopelessness of his efforts, the apeman gave up, and returned to the outcrop at the top of the hill.

The sun was up, and the others were now awake. They foraged for food among the shrubbery in a ravine on the north face of the kopje. There was little new growth. Most greens leaves were mature and covered in the fine dust that accumulates with no rain. The apeman and his group fed until midmorning and then headed out onto the veldt. As they reached the burnt land, the calls of two group members who stayed behind prompted them to return.

The apeman brought up the rear, as the group ascended the kopje to the place where they spent the night. The sun's position was just right to illuminate the cave floor. The apeman looked down into the cave, just as he had done many years ago. Lying motionless with her limbs akimbo on the cave floor was his daughter. He called out to her, but there was no reply. The group remained there for the rest of the morning. Periodically a member of the group would call to the body below, but it never stirred. By late afternoon the group was in a stand of trees growing adjacent to a water-saturated glade. The apeman unearthed a tuber from the base of a cabbage tree and ascended the tree. planning to sleep in it for the night. The trees by the glade had plenty of water and thus new leaf growth. The apeman and his group would stay there until the coming rains made food in the veldt plentiful again.

AUSTRALOPITHECUS AFRICANUS

Skull, Teeth, and Diet

Australopithecus africanus exhibits many human-like skull and tooth features. As in humans, the face is relatively flat and vertically disposed, and the jaw is relatively short. The neck musculature attaches low on the skull, and the foramen magnum (the hole at the skull base for spinal column passage) is positioned forward and oriented downwards more so than in African apes.

The *Australopithecus africanus* canines 1) lack honed distal edges, are not associated with honing lower premolars, and in most cases lack gaps (i.e., diastemas) between them and adjacent teeth, 2) are small and do not protrude past the tooth (incisors) chewing surfaces, wearing mainly at the tip, and 3) appear to show limited differences in crown development between males and females. Notably the first milk molar shows the multicusped condition common to humans and *Paranthropus,* which is absent in great apes. Because many Miocene (from 23.5 million to 5.3 million years ago) and Pliocene (from 5.3 million to 1.8 million years ago) fossil apes, some of which are not exclusive human ancestors (e.g., *Ouranopithecus* and *Oreopithecus*) may show these tooth features, it is likely that they reflect a generalized condition common to human and African ape ancestors.

The *Australopithecus africanus* skull also appears to be generalized. Its 400 cc brain volume is close to the mean values shown by great apes. It lacks the strong development of the chewing musculature and marked canine and front tooth reduction characteristic of *Paranthropus* (see *Paranthropus robustus/crassidens,* Skull, Teeth, and Diet). In contrast to the flat wearing *Paranthropus* molars, the *A. africanus* molars have gutter-like wear reflecting more of a forward-backward movement of the jaw during chewing, as opposed to the specialized rotatory jaw movement shown by *Paranthropus.*

As in *Paranthropus,* the *Australopithecus africanus* teeth are large relative to body size. *Australopithecus africanus* relative molar size and simple molar cusp pattern, without specialized shearing crests, finds an analogy with the baboon molars, suggesting it had a relatively generalized diet, which was probably low in nutritional content and had to be ingested in bulk. Enamel micro-wear analysis of its molars suggests it ate leaves and possibly seeds, and its diet was much less abrasive than that of *Paranthropus.*

Larger and more developed front teeth, when compared to *Paranthropus,* and a forward-backward movement of the jaw indicate *Australopithecus africanus* front teeth were more important in food processing, e.g., husking, stripping, nipping, and cutting than for *Paranthropus.* Suggestions based on carbon isotope analyses of Makapansgat molars that the *A. africanus* diet did not differ from that of *Paranthropus* probably reflect the presence of *Paranthropus* at Makapansgat and the difficulty of assigning fragmentary fossil remains (in this case unassociated badly-worn molars) to their correct group. *Australopithecus africanus* probably ate many of the same foods consumed by baboons, vervets, and samango monkeys presently inhabiting temperate zones in South Africa, showing differences in foods eaten according to season. During the dry season when food was scarce, *A. africanus* must have occasionally fed on meat and eggs. Herbs, fruits, flowers, and pith were most commonly consumed in the spring and summer when they were most available. Leaves, tubers and roots, lichen, bark, and seeds would have been consumed throughout the year, and depended on when spring and summer foods were scarce. Insects (mainly grasshoppers, termites, and grubs) and honey would have been fed on whenever available.

Skeleton, Gait, and Posture

Even though the most complete Sterkfontein skeleton known (StW 573) has not yet been fully removed from breccia, the rich collection of *Australopithecus africanus* skeletal parts already unearthed gives an accurate indication of its body proportions and movement capabilities. Upper limb long bone skeletal length appears to be approximately 85 percent that of the lower limb, a value halfway between humans (76 percent) and pygmy chimpanzees (92 percent), and similar to that of ground-living quadrupeds such as baboons. As in gorillas and humans, the forearm appears to be shorter than the arm; and, as in all members of the great ape–human lineage, the lower leg is shorter than the thigh. Relative to body weight, the pelvis is much longer than it is in humans.

Although in most respects the hand is human-like, it appears to be larger relative to body size than the human hand. Length to mid-shaft diameter ratios of individual hand and foot bones (i.e., metacarpals, metatarsals, and manual and pedal phalanges) indicate relatively longer fingers and toes than seen in humans. The foot bones (StW 532, nicknamed Little Foot) reflect a much more mobile foot that lacked fixed arches (longitudinal and transverse) and had a greater degree of big toe divergence than do modern humans. Two lower backbones in contact with the pelvis (Sts 14 and StW 431) show a cross-sectional joint area that is small relative to the thigh bone and arm bone shaft circumferences, suggesting that

Composite reconstruction of the skull of an adult *Australopithecus africanus* based primarily on Sts 5 ("Mrs. Ples") discovered by Robert Broom and John Robinson in 1947, and the jaws of Sts 52.

Australopithecus africanus upper body weight was not supported solely by the hindlimbs. The angle the lower backbone makes with the pelvis indicates *A. africanus* probably lacked the lower back curvature (lumbar lordosis; *see Homo sapiens*) associated with routine two-legged movements and stance. It is highly unlikely that A. *africanus* spent all of its time on two legs. Two-legged stance and movements were probably used mainly during feeding. Such behaviors would best explain the degree of hip joint extension, suggested by the Sts 14 and StW 431 hip joints, a larger and more rigid pelvis/lower back joint than in the A.L. 288-1 *Australopithecus afarensis* skeleton (Lucy), and the arguable presence of a human-like carrying angle in two thigh bone fragments (TM 1513 and Sts 34).

Overall, the *Australopithecus africanus* body proportions and joint movement capabilities are most in accord with a four-legged ground moving early hominid that still retained the ability to climb trees and spent considerable time standing on two

Australopithecus africanus

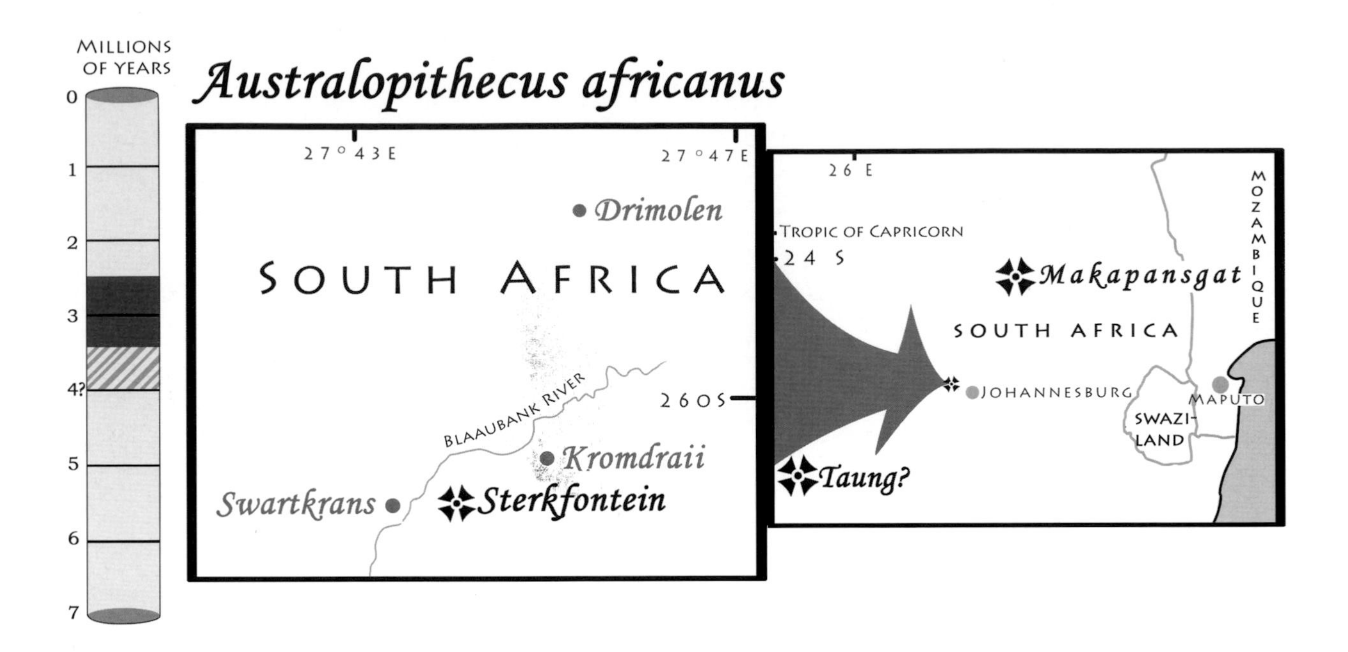

legs and in erect trunk postures during feeding. Its small size (25 kg to 30 kg) and proportions would have made it a more adept tree climber than either gorillas or modern humans. Given its taste for fruits and leaves, it would have slept in trees whenever these were available. At least in the subtropical climates it inhabited, it must have occasionally slept or rested on the ground.

Fossil Sites and Possible Range

Australopithecus africanus is found in two breccia bearing cave deposits at Sterkfontein and Makapansgat in the Transvaal dolomites. The Sterkfontein cave deposits are found approximately thirty-five kilometers west of Johannesburg, South Africa, halfway between the Swartkrans and Kromdraii B caves. The Makapansgat Cave deposits are found 16 km east-northeast of Potgetiersrus at approximately two degrees latitude north of Sterkfontein. Although fossil teeth from cave deposits at Coopers B and Gladysvale could possibly represent *A. africanus*, without more complete and associated remains the designation of these fossils is uncertain.

Most *Australopithecus africanus* fossils come from Sterkfontein Member 4, the second most recent of the five members into which the deposit is divided. Because the Sterkfontein caves have been excavated almost continuously since 1936 and for approximately ten years prior to that served amateur collectors, they are the main source of our knowledge on *A. africanus*. Most remains are fragmentary and distorted from the fossilization process, but a number of complete or nearly complete skulls (Sts 5, Sts 71, and StW 505) and associated skeletons (Sts 14 and StW 438) have surfaced. In varying condition and state of completeness, nearly all *A. africanus* skeletal elements are represented.

Despite decades of systematic excavations carried out in Makapansgat's bone rich gray marl, this cave deposit has yielded relatively few *Australopithecus* fossils. *Australopithecus* may not have been as relatively common in the Makapansgat area as it was in Sterkfontein.

The Taung Cave Deposits and the Holotype Specimen of *Australopithecus africanus*

The *Australopithecus africanus* holotype specimen, a child's skull, is alleged to come from breccia bearing cave deposits at the old Buxton Norlim limestone quarry at Taung. The exact location within the deposit where the skull was found, however, has never been identified and is believed to have been lost to mining operations. Because 1) no other *A. africanus* remains were ever found in or around the Buxton Norlim Quarry, 2) the known cave deposits in and around the quarry are all at least a million years younger than other known *A. africanus* deposits, and 3) the depositional climate the Taung breccias indicate is much drier than that indicated by the breccia from which this child's skull supposedly was extracted, there is considerable uncertainty as to where this holotype specimen actually came from.

Some of the Buxton Norlim Quarry miners also worked in the Sterkfontein area (e.g., George Barlow and Knowlan). It is possible that the breccia block containing the child's skull was a keepsake transported by one of the miners from a prior mining site in the Sterkfontein Valley to Taung. When a request was made to send fossils to Raymond Dart (*see* below, Historical Notes), it is not inconceivable that these keepsakes were sent along with fossils found in the Taung quarries. Otherwise, there is no biological explanation as to how *Australopithecus africanus*, a fossil species that disappeared sometime before two million years ago in the Sterkfontein and Makapansgat area, could pop up virtually unchanged one million years later at Taung in what was certainly a much drier climate.

Age

None of the *Australopithecus africanus* deposits have so far yielded absolute dates. Biostratigraphic estimates suggest Sterkontein Member 4, from where most individuals come, is approximately two and a half million years old. Members 1-3 are somewhat older, possibly dating back in time to 3.5 million years. A recent 4 million-year date for the earliest Sterkfontein Member based on new absolute dating techniques has yet to be verified. The Makapansgat deposits are clearly contemporaneous with Sterkfontein Member 4, but also extend to more recent times, overlapping with Kromdraii A and B and possibly Sterkfontein Member 5. The oldest deposits at Makapansgat are probably younger than the oldest deposits at Sterkfontein, although this is not clear.

Two independent studies based on relative dating methods have both reached the conclusion that the known Taung Cave deposits are around one million years old. A Taung origin for the holotype child's skull, therefore, seems unlikely given the 1) large age differences between Sterkfontein and the known Taung deposits, 2) strong similarities between some Sterkfontein fossils and the *Australopithecus africanus* holotype, suggesting membership in the same species, and 3) the inability to find a deposit at Taung from which the *A. africanus* holotype could have originated.

Tool Use

Until the early nineteen sixties, it was believed that *Australopithecus africanus* had made and used the stone tools found at Sterkfontein. This belief was based on an erroneous association of the Sterkfontein Member 5 stone tools with Sterkfontein Member 4 *A. africanus* fossils. As of yet, stone tools have never been found in the Sterkfontein *A. africanus* deposits. Considering that both chimpanzees and orangutans are known to fashion tools, *A. africanus* probably also made and used tools. At least in the Sterkfontein area, whatever tools they may have used were very rarely or never made of stone. In Makapansgat, stone tools were reputedly found in a layer between *A. africanus* fossils, suggesting to some that it was a toolmaker and user. The fossils found, however, are too fragmentary to have a definite *A. africanus* designation. At the lower Makapansgat levels where most fossils, including most *A. africanus* fossils, are found, there are no stone tools.

Appearance

Considering incident sunlight and radiation at high elevations, *Australopithecus africanus*, like other African early hominids, probably had dark hair and facial skin, which would protect against UV rays. No evidence presently exists as to the length and density of *A. africanus* body hair.

Growth and Development

Rate of tooth enamel deposition when contrasted to tooth eruption suggests *Australopithecus africanus* had maturation periods more like those of living

African apes than those of modern humans. Unfortunately, duration of great ape developmental stages are well documented only for captive animals. Preliminary data from free-ranging African apes shows that at least for longevity and sexual maturation there is a large range of variation and a broad overlap with humans.

Differences Between Males and Females and Social Systems

Differences in absolute size between skulls from different individuals, e.g., the Sts 19 and StW 505 skulls, indicate that males may have been much larger than females. Such male to female size differences suggest a harem-type social system in which groups are composed of one or two adult males and several reproductive females and their offspring. The relatively harsh environmental conditions suggested by the present day climate are also in accord with a harem-type social system. Unfortunately, there are no clear anatomical indicators that enable small males to be distinguished from large females, so that *Australopithecus africanus* size dimorphism and thus a postulated harem-type social system is not certain.

Animals and Habitats

In general, most mammals found in the same deposits as *Australopithecus africanus* are more or less modern in aspect. Monkeys and carnivores are the exception. All four monkey species known from Sterkfontein Member 4 belong to two genera (*Parapapio* and *Cercopithecoides)* that are presently extinct. There are no gelada baboons or baboons of modern aspect as appear in Swartkrans. Half of the carnivore genera present at Sterkfontein Member 4—*Megantereon* (the saber-tooth), *Dinofelis* (the false saber-tooth), and *Chasmoporthetes* (the hunting hyena)—are also presently extinct. In contrast, only three (*Parmularius, Megalotragus*, and *Makapania)* of the fifteen ungulate (hoofed plant-eaters) genera found at Sterkfontein Member 4 are extinct. The disproportionate extinction of monkeys and carnivores suggests competition with either early hominids or later *Homo*. Notably, monkeys are three times more common than *A. africanus* at Sterkfontein Member 4, while *Paranthropus* at Swartkrans exists at similar or higher densities than do monkeys. It may very well be that the increase in numbers of *Paranthropus* was directly associated with the disappearance of endemic subtropical monkeys in southern Africa. Gelada baboon presence in the Makapansgat deposits is probably a reflection that the deposits span to a more recent age than Sterkfontein Member 4. It remains to be shown whether gelada baboons were contemporaneous with *A. africanus*. All animals found indicate *A. africanus* inhabited grasslands with woodlands close to watrer sources, resembling, more or less, the habitats found in the area today.

Climate

Found at approximately fifteen hundred meters above sea level and twenty-six degrees south of the Equator, the Sterkfontein caves presently enjoy a temperate climate. Winters (June-August) are marked by low temperatures with daily minimums around or below the freezing mark and very little rainfall. The area receives an average of only 710 mm of annual rainfall, mostly occurring during summer and late spring (November-February 450 mm, 63 percent of the total yearly average). The fifteen-hundred-meter-high elevation results in large fluctuations in daily temperatures that commonly drop 14° C during winter nights. Light snow may occur occasionally in late August or early September. With an average annual rainfall of 580 mm, Makapansgat is presently drier than the Sterkfontein area. Surprisingly, at 100 m lower elevation and two degrees closer to the Equator, Makapansgat is on average 4° C colder than

Peering out from behind the foliage, a female *Australopithecus africanus* keeps a pensive eye over her family for predators.

Reconstruction of Raymond Dart's *Australopithecus africanus* Taung Child skull.

Sterkfontein. With approximately four hundred millimeters of rain annually, Taung is presently much drier than both Sterkfontein and Makapansgat.

There is no evidence to suggest the climate was markedly different when *Australopithecus africanus* lived in these areas than it is at present. Cave sediment chert (a rock chemically precipitated in ocean water) to quartz ratios and grain angularity have been shown to reflect differences in local rainfall and suggest that the Sterkfontein deposits began to accrue when rainfall was similar to what it is today. The chert accumulated in cave sediments comes from within the dolomite, while the majority of quartz in the sediments arrives as dust carried by the wind. High rainfall dissolves dolomite faster, releasing a greater amount of chert into cave sediments, but clears the air of wind carried dust, reducing the amount of quartz in cave sediment. Changes in rainfall over time are thus tracked by changes in the ratio of chert to quartz. In addition, small particles carried a long distance by the wind are abraded (taking on a rounded shape), while those travelling short distances are more angular. Increasing rainfall, therefore, decreases wind-born particales, increasing the angularity of small particles incorporated into cave sediments. As the Sterkfontein caves filled, both these ratios show, the climate became progressively more arid, reaching a low point of approximately five hundred millimeters of rain annually, before returning to today's rainfall levels by the time Member 4 was fully formed. During deposition, Makapansgat also appears to have undergone a dry period. In both areas, it is possible that a dry period may have been associated with slightly higher temperatures than occur today.

Classification

In a February 1925 letter to *Nature*, Raymond Dart coined the binomial *Australopithecus africanus* for the skull of the Taung Child, making it the holotype specimen of his new species. In a letter to *Nature* in 1936, Robert Broom assigned the first hominid fossil found at Sterkfontein (the TM 1511 partial skull) to the same genus as the Taung Child. His report on a molar in the same journal the following year formalized the species designation for all Sterkfontein early hominids as *Australopithecus transvaalensis*. A year later (1938), Broom elevated the Sterkfontein species to a new genus *Pleisianthropus transvaalensis*, claiming a newly found lower jaw from a child (TM 1516) was distinct from the *A. africanus* holotype specimen. In 1948, Dart gave the name *Australopithecus prometheus* to a fragmentary skull base found by James Kitching at Makapansgat. All the Sterkfontein and Makapansgat early hominid fossils were later assigned by various paleoanthropologists to *A. africanus*, a practice that has since been followed.

Unaware of his approaching fate from above, a juvenile *Australopithecus africanus* greets a new morning two and a half million years ago.

The following facts, however, confound *Australopithecus africanus* classification so that it is far from cut-and-dry: 1) Sterkfontein Member 4 appears to have at least two different kinds of early hominid fossils, i.e., one with a long palate, parallel tooth rows, and a long skull base (e.g., the StW 258 and Sts 5 skulls), and one with a short palate, tooth rows set in an arc, and a short skull base (e.g., the Sts 53 and Sts 19 skulls); 2) Makapansgat early hominid deposits, although contemporaneous with Sterkfontein Member 4, also span to more recent times, possibly yielding early *Homo* (the MLD 7 and MLD 25 juvenile pelvis fragments) and/or *Paranthropus* (the MLD 2 lower jaw) fossils; 3) the first permanent molar is the only definite adult feature of the *A. africanus* holotype (the Taung Child), and no adult *A. africanus* has yet been found at Taung; and 4) the *A. africanus* holotype specimen comes from an unknown deposit. Because the oldest deposits at Taung are more or less contemporaneous with the youngest *Paranthropus* bearing deposits at Swartkrans, some paleoanthropologists have even entertained the possibility that the Taung Child may represent *Paranthropus*, throwing a further monkey wrench into *A. africanus* classification. A number of the Taung Child's tooth and skull features, however, show that it is probably the same species as the long-jawed hominid at Sterkfontein and clearly distinctive from *Paranthropus*.

Australopithecus transvaalensis is an available name for the short-jawed species at Sterkfontein. Although some Makapansgat remains (the MLD 1 fragmentary skull base and MLD 37/38 skull) may represent *Australopithecus africanus*, it is clear that others do not (the MLD 2 lower jaw and the MLD 7 and MLD 25 juvenile pelvis fragments). The Makapansgat fossils are too few and fragmentary to resolve classification at the species level. Future finds at this site may very well demonstrate that Dart's *Australopithecus prometheus* is a distinctive species.

Historical Notes

Australopithecus africanus was first found by Raymond Dart in November 1924, in a fossil collection the then manager of the Taung Buxton Norlim Quarry, A. Spiers, had sent to him. M. deBruyn, an old miner working for Spiers, had supposedly gathered the fossil collection at the Taung quarry. The binomial name Dart chose for his new species literally translates to mean "Southern ape of Africa." The fossil consists of the nearly complete face of a child, including the first permanent molars, a full set of upper and lower milk teeth, and a mold of the internal surface of the braincase. It is currently housed in the Anatomy Department at University of Witwatersrand Medical School. Broom is credited with finding the first early hominid at Sterkfontein (TM 1511) in 1936. Two years later he designated it as the holotype specimen of a new genus and species, *Plesianthropus transvaalensis*. Broom chose the Greek "plesios," meaning "near," as a suffix for the genus name to reflect his belief that it was close to human ancestry. The holotype specimen is an adult, consisting of a partial face, forehead, a palate including the cheek teeth, and a natural cast of the internal surface of the braincase. It is housed in the Transvaal Museum, Pretoria. James Kitching is credited with finding the first early hominid at Makapansgat, the MLD 1 skull base. It is presently housed in the Anatomy Department at University of Witwatersrand Medical School. Dart described the fossil in 1948 as the holotype specimen of a new species, *Australopithecus prometheus*, naming it after the Greek god who gave fire to man. This name reflects Dart's prior belief that the occurrence of charred bone at Makapansgat showed controlled use of fire by protohumans. Dart also noted that "Prometheus" in Greek means "forethought," something he believed this early hominid possessed.

TO LIVE AND DIE ON THE HIGH VELDT

From a corner of the cave's entrance came a baby's cry. A mother proudly clutched her newborn infant in her arms while he suckled. Her oldest daughter looked on, impatiently, waiting to hold her youngest brother. Before sunset, everyone in the group had inspected the new member. They held him and fussed over him. Some ran a finger along his face and brow, and others smacked him with their lips. That night, the wind blew hard, forcing the group to sleep well behind the cave's entrance, but not so far behind as to obstruct a view to the outside.

The next morning when the sun rose, the high veldt was blanketed in snow. Squatting at the cave entrance, the apemen looked out at the world below. A translucent ice sheet covered the pond at the glade center. A mixed herd of wildebeests, zebras, and springbok had broken through the ice and were taking turns to drink. Guinea fowl were pacing back and forth in their empty tracks, kicking up snow and ice, and overturning pebbles to find food. On the opposite side of the glade, the barely recognizable, snow covered carcass of a stray hippo still attracted a lone jackal and some vultures.

By mid-day the sky had cleared and the air laid still. Only a few small blotches of frozen white interspersed in crevices and shadows served as testament to the snow that had covered the veldt earlier that day. The sun's heat was in stark contrast with the cool air. The sun's rays hitting warmed the apemen, prompting them to leave the cave entrance to forage. The group ventured to the pond's edge filling their mouths with rhizomes and small bulbs they dug up with thumb and forefinger from the soft and muddy shore. The mother stayed behind, clutching her newborn infant against her body and holding an antelope's horn in her free hand.

She waddled a short distance to a patch of wild lilies that grew adjacent to the cave entrance. Stabbing at the hard rocky soil with the horn, she extracted a large bulb that she shared with her daughter and her 6-year old son. All the while her newborn infant suckled at her breast.

With the first rains of November the brown and burned grasses of the high veldt gave way to an iridescent green of new growth. New leaves on trees growing from water-saturated holes in the dolomite, dotted the landscape with clumps of bright foliage. A gelada baboon troops that congregated around the glade in the dry season had long left to forage in new growth that stretched out endlessly into the horizon. The apemen moved along a stream bank, stopping every so often to feed on fruits or new foliage. Lagging behind the main group, the daughter and her brother fed on small berries in the crown of a tree. Leaf-eating monkeys fed next to them, warily keeping a safe distance.

Rain clouds that had been lurking in the sky since mid morning grew to cover the sun and darken the veldt. In the distance, a tortuous lightning bolt fell on the horizon exploding in a flash of light. The deafening boom that immediately followed startled the apemen. Strong wind gusts arose, pulling hard at the trees, bending crowns and breaking branches. Nervous and disoriented the group scattered in search of cover. The oldest daughter and her younger brother found shelter in a small hole at the base of a large fig tree near a small stream. The earth in the hole was soft and comfortable. The two huddled together, shuddering with each flash of light and the loud explosion that followed. The rain poured from the sky and continued to do so almost until sunset. When deluge finally ended, the small meandering stream had turned into a churning river. Dry but groggy from the afternoon's inactivity, the apemen, far ahead of the two, slowly headed back to the caves by the glade, pausing occasionally to feed.

When the daughter and son found a way across the stream and reached the glade, it was almost dark. They could barely see the others silhouetted against the cave entrance. Without warning, a leopard sprung from thickets and clamped its jaws around the son's throat. Screaming and swinging a thorny branch, the daughter tried to force the leopard to relinquish its hold. A muffled shriek and in seconds the son's flailing body went limp. Others from the group appeared, brandishing branches and hurling rocks. Harassed, but not the hurt, the leopard took its prey by the head and ran of into the darkness.

In the span of twelve years, the mother had given birth four times. The oldest a daughter and the youngest, a twelve-week-old infant, were still alive. She lost one child at birth and now she had lost one to a leopard. Nearly an adult, her daughter would soon leave to join another group and find a mate. The mother was too preoccupied with her new infant son to think about her recent loss, or the impending loss of her daughter and closest ally.

PARANTHROPUS ROBUSTUS/CRASSIDENS

Skull, Teeth, and Diet

The *Paranthropus robustus* head, especially the teeth and chewing muscle attachment areas, appear large relative to body size. The teeth are characterized by large flat wearing premolars and molars and relatively small peg-like front teeth (incisors). As in humans, and in contrast to living African apes, the canines 1) lack honed distal edges and are not associated with tooth row gaps (i.e., diastemas) or honing lower premolars, 2) are small and do not protrude past the front tooth chewing surface, wearing mainly at the tip, and 3) do not appear to show marked differences in crown development between males and females. Premolars are molarized, showing chewing surfaces approximating those of the molars in area. In further contrast to African apes, the face is relatively flat and vertically disposed. The neck musculature attaches low on the skull, and the foramen magnum is positioned forward and oriented downwards, suggestive of possible two-legged behaviors or habitually erect postures. There is, however, overlap between humans and African apes in foramen magnum position and neck muscle attachment disposition so that these features cannot be used to conclusively determine gait and posture. Overall the morphology of *Paranthropus* is not exclusively human-like nor in itself shows routine two-legged movement or posture as has been argued by some. At 515 cm, its brain volume is about 100 cm larger than the chimpanzee average.

The marked cheek tooth and chewing musculature development in *Paranthropus* reflects a diet ingested in bulk that was low in nutritional content. Analogous specializations are seen in herbivorous mammals that consume large quantities of roughage and spend a large portion of the day eating. Small front teeth and canines suggest a limited role in preparing food items for molar grinding. The items fed on may have been small in size or processed down to size largely by the hands. As do gorillas, *Paranthropus* probably ate from a wide range of plant foods, including fruits, and consumed many of the same plants favored by baboons, verbets, and samango monkeys presently inhabiting temperate zones in South Africa. In further analogy to these primates, *Paranthropus* must have shown seasonal differences in diet, feeding on bark, lichens, roots and tubers (including those of grasses), and possibly insects and small vertebrates in late fall and winter, during the dry season when fruit and foliage were not abundant. Foliage, herbs, pith, and fruits were most abundantly consumed in late spring and summer.

Enamel micro-wear analysis of *Paranthropus* molars shows evidence that it ate hard abrasive foods. Carbon isotope analyses of its teeth, based on accumulation of the heavier C^{13} isotope in grasses but not in other plants, indicate that *Paranthropus* at all ages fed at least partially on grasses. Interpretation that a relatively high C^{13} isotope presence indicates *Paranthropus* preyed on animals that fed on grasses is not convincing considering the isotope's abundance in the deposit and Paranthropus' chewing specializations that suggest a committed plant-eater.

Composite reconstruction of the skull of *Paranthropus robustus* based on the SK 46, SK 48, and SK 12a and b skulls from Swartkrans limestone caves.

Skeleton, Gait, and Posture

With the exception of the skull the remainder of the *Paranthropus* skeleton is poorly known. With no associated long bones or long bone fragments able to provide length estimates, its body proportions are unknown. Most long bone joints either are not represented in the remains (shoulder and knee joints) or include too few elements (ankle and one damaged talus, TM 1517; a wrist and a forearm bone, SKW 3602; and a wrist bone SKW 3498) to provide insights into joint movement capabilities.

What little is present, however, is strikingly human-like, begging the question as to whether we can consistently identify *Paranthropus* fossil fragments as separate and distinct from those of *Homo*. Length to mid-shaft diameter ratios of individual hand and foot bones (i.e., metacarpals, metatarsals, and manual and pedal phalanges) indicate human-like finger and toe lengths. Relative thumb size is also human-like and strongly contrasts with that of chimpanzees. Two damaged pelvises (Sk 50 and Sk 3155) are also seemingly human-like. A backbone with a joint surface for the pelvis (Sk 3981b), however, shows a joint orientation that indicates *Paranthropus* probably lacked the lower back curvature (lumbar lordosis) associated with routine human two-legged stance and movements. The relatively small joint surface area of this backbone suggests the upper body weight of *Paranthropus* was not supported solely by it's hindlimbs. Two fragmentary thigh bones (Sk 87 and Sk 92) lack evidence of a human hip joint locking mechanism.

All of the above suggests *Paranthropus* was not a habitual biped and probably engaged in some four-legged behaviors. Two-legged movements were probably used for feeding, when moving between feeding sites, and when walking along wet stream banks. Four-legged movements probably occurred in heavy underbrush, on unstable or slippery substrates, or when climbing steep inclines. Sitting postures with an erect trunk may have been common when gathering food from the ground. Although its short fingers and toes clearly indicate *Paranthropus* was mainly terrestrial in its habits, its relatively small body size (between 25 kg and 35 kg) would have made it more adept in trees than either gorillas or humans. Given a relative scarcity of large trees in its habitat, *Paranthropus* probably slept in rocky outcrops or in cave entrances, as presently done by baboons inhabiting these areas. It most likely returned to those sleeping sites that over time proved safest.

Fossil Sites and Possible Range

Paranthropus robustus is known from three breccia bearing cave deposits—Kromdraii B, Swartkrans, and Drimolen—in the Transvaal dolomites 35 km west of Johannesburg, South Africa. The three deposits are relatively close to each other. The Kromdraii B deposit is 2.8 km east of Swartkrans, and Drimolen is 4.3 km north of Kromdraii. Isolated teeth from other nearby breccia bearing cave deposits in these same dolomites, i.e., Gondolin, Gladysvale, Coopers B, and the Sterkfontein Member 5 deposit, may also be

Paranthropus robustus/crassidens

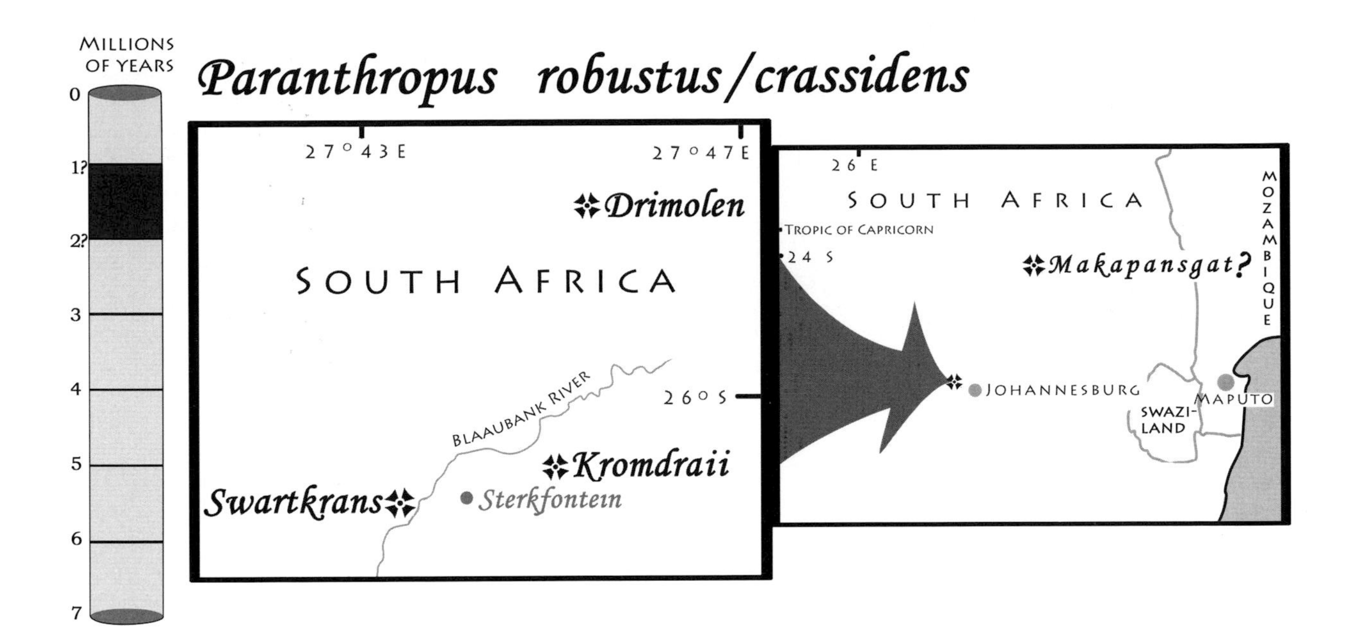

referred to *Paranthropus*. These remains, however, are too fragmentary to provide unequivocal classification. A juvenile jaw MLD 2 and other more fragmentary Makapansgat fossils may also represent *Paranthropus*, but given a number of distinctive features their classification as to species is uncertain.

At all sites, *Paranthropus* is known mostly from dental remains and skulls, some of which are nearly complete (the Sk 48 skull and the Drimolen DNH 7 female skull), although somewhat damaged. Because most fossil individuals come from Swartkrans Members 1-3, where *Paranthropus* is one of the most common mammals in the deposit, most of what we know about *Paranthropus* is based on Swartkrans. Unfortunately, the occurrence of *Homo* in most *Paranthropus* bearing deposits (with the exception of Kromdraii B and Swartkrans Member 3) and the close anatomical similarities between the two create considerable uncertainty as to which fossils belong to which population, especially when fossils are unassociated or fragmentary. Considering *Paranthropus* occurs in these deposits at much higher frequencies than *Homo*, there is a greater likelihood that non-diagnostic fossils that cannot be assigned with certainty to either population belong to *Paranthropus*.

Age

None of the deposits have so far yielded absolute dates. Biostratigraphic estimates suggest that the three members the Swartkrans hominid deposits represent an interval between 1 million and 1.8 million years ago. Kromdraii B is likely to be somewhat older than Swartkrans, extending as far back as 2 million years. What is so far known suggests that Drimolen represents a time span that fully overlaps Swartkrans Members 1-3.

Tools and Fire

In order of decreasing age, Kromdraii B and then Swartkrans Members 1-3 show an increasing density of bone and stone tools. Classically it has been assumed that these tools were the sole provenance of *Homo*. The absence of *Homo* in the one Swartkrans member (Member 3) where most bone tools are found, however, tends to suggest that *Paranthropus* also used tools.

Heavy tip wear on tools indicates to some that they were used to dig, possibly to extract tubers. Other studies have emphasized striated wear along the tool shaft to argue that these tools were used to fish for termites. Neither function is mutually exclusive. In Swartkrans Member 3, burned-bone

at temperatures too high (300° to 500° C) to be associated with grass fires indicates fire use. Bones also from Member 3 with stone tool cut marks that indicate defleshing suggest meat-eating. Considering that *Paranthropus* used tools, evidence of fire use and meat-eating in the Swartkrans Member 3 deposits is as likely attributable to *Paranthropus* as to *Homo,* further obscuring the differences between these two populations.

Appearance

Considering incident sunlight and radiation at high elevations, *Paranthropus* probably had dark hair and facial skin, which would protect against UV rays. Because external ear size in primates in part reflects ability to dissipate head heat, a small ear size is expected in an animal that given a low nutrient diet probably had a low metabolism. No evidence presently exists as to the length and density of *Paranthropus* body hair. Body hair reduction and loss is not a feature unique to modern humans. Other mammals have also lost their hair either as a response to body heat control and/or to avoid external parasites (e.g., lice, fleas, and ticks). The possibility that *Paranthropus* may have slept together in large groups to keep warm on cold nights may have required hair loss to reduce the rate of parasite transmission.

Growth and Development

Rate of tooth enamel deposition when contrasted to tooth eruption suggests *Paranthropus* had maturation periods more like those of African apes than those of humans. Preliminary data from free-ranging African apes shows that at least for longevity and sexual maturation there is a large range of variation and a broad overlap with humans.

Lost in a patchwork of shrinking forests and expanding savannahs, *Paranthropus robustus* chomps on his favorite leaves while lounging in an oasis of dense vegetation along a stream in the Transvaal.

Differences Between Males and Females and Social Systems

(*See Australopithecus africanus.*)

Animals and Habitats

In general, mammals found in the same deposits as *Paranthropus* are modern in aspect. Old World monkeys and large cats, however, are characterized by a high proportion of genera that are now extinct. Representing five different genera (*Theropithecus, Papio, Dinopithecus, Parapapio,* and *Cercopithecoides*) and varied diets (a grass-eater, three omnivores/fruit-eaters, and leaf-eaters), these Old World monkeys indicate a primate community that may have fed and traveled together with *Paranthropus* to reduce the risk of predation. Occasionally these monkeys may have preyed on *Paranthropus* infants or have been the prey of *Paranthropus.* The extinction of the majority of these primates and of false saber-tooths (*Dinofelis*) by Swartkrans Member 3 times suggests some degree of competition with either *Paranthropus* or *Homo.*

Considering a relatively low rainfall, it is unlikely that there would have been a high density of mammals in the area as may exist in tropical grassy plains or in artificially watered national parks. What mammals were present, however, were likely to congregate around water sources in the dry season. Notably, both Drimolen and Swartkrans are caves associated with water bearing glades where animal congregations would have occurred.

Paranthropus robustus/crassidens as Prey

Evidence that *Paranthropus* was hunted by leopards comes from two indentations on a juvenile skullcap (Sk 54). The size of the indentations and the distance between them correspond to leopard canines.

Climate

There is no evidence that suggests the climate was any different when *Paranthropus* lived in the area than it is at present. Quartz to chert ratios and angularity of sediment grains, however, indicate that the Kromdraii B deposits may have been formed during somewhat wetter times (1000 mm of rain annually) than exist today. The occurrence of springbok, an antelope that can inhabit frost cover areas, in the same deposits *Paranthropus* is found is consistent with a climate as cold or colder than it is at present.

Classification

The Kromdraii hominid remains were originally referred to *Paranthropus robustus* and the Swartkrans hominid remains to *Paranthropus crassidens.* The fragmentary nature and small sample size (six individuals) of the Kromdraii remains confound their separate classification. Most paleoanthropologists have considered the two the same species, *Paranthropus robustus.* The likelihood of specific differences between the two samples, however, is supported by the supposed earlier age of Kromdraii and a tooth size and shape that seems in many ways intermediate between the Sterkfontein and Swartkrans hominids. Drimolen early hominids are most similar to those from Swartkrans, and thus referable to *Paranthropus crassidens.* This specific designation could very well change when more Drimolen specimens are uncovered.

Historical Notes

Gert Terblanche, a schoolboy living close to Kromdraii, is credited for first unearthing this early hominid's remains. At the time he gave part of the fossil to George Barlow, the manager of the Sterkfontein lime works, who in June of 1938 brought it to the attention of the renowned fossil collector Robert Broom. Aware of its significance, Broom contacted Terblanche and returned with him to Kromdraii B to unearth the rest of the fossil.

The fossil consists of a damaged skull, half a jaw bone, a damaged anklebone (talus), and a partial elbow joint (corresponding fragments of the ulna and humerus). A partial and damaged hand recovered at the same time and thought to belong to the same individual has been shown to belong to a baboon. The fossil is in the collections of the Transvaal Museum Pretoria cataloged as TM 1517.

In a 1938 note to the scientific journal *Nature*, Broom coined the name "*Paranthropus robustus*" for the Kromdraii hominid, assigning it to a new genus and species. Broom's use of the Greek root "para" in the fossil's generic name reflects his opinion as to its phylogenetic position, which he believed to be on a side branch of the human lineage. The species name "*robustus*" refers to its robust teeth and skull.

In a note to *Nature* in 1949, Broom used the name *Paranthropus crassidens* to refer to fossils he and John Robinson had unearthed during their 1948-1949 excavations at Swartkrans. Broom believed these fossils to differ enough from Kromdraii to merit a separate species. The species name "*crassidens*" is from the Latin word for "thick" and describes this fossil's large teeth and jaws. Broom designated a jaw from the Member 1 deposit, now cataloged as Sk 6 in the Transvaal Museum, as the holotype specimen of his new species.

The first fossil hominid from Drimolen was found in July of 1992 when Esteban Sarmiento and Andre Keyser identified a molar, which a technician was digging out of a breccia pinnacle, as the upper left molar of *Paranthropus.* Since then, Drimolen has yielded one of the most complete early hominid skulls ever found.

BACK TO THE EAST AFRICAN GREAT RIFT VALLEY AND THE APPEARANCE OF *HOMO*

Homo rudolfensis • *Homo habilis* • *Paranthropus boisei* • *Homo ergaster*

A number of incomplete hominid fossils, mainly teeth and damaged jaws from *Australopithecus* and *Paranthropus* bearing deposits, have been either tentatively assigned to *Homo* or assigned to *Homo* without designating a species. These fossils come from Sterkfontein Member 4, South Africa; west Lake Malawi Chiwondo beds, Malawi; Tugen Hills *(see Paranthropus aethiopicus)* and west Lake Turkana Lokalalei Member, Nachukui Formation, Kenya; Hadar upper Kada Hadar Member, Ethiopia; and Shungura Members B-G, Omo Basin, Ethiopia. Most of these fossils span a time interval between 2.3 million and 2.6 million years ago, with the exception of those from Omo, with dates of 2.0 million to 3.0 million years, and Sterkfontein, with less certain dates of 2.5 million to 3.0 million years. Their problematic classification stems from their incompleteness. They lack features that are either exclusive to any known *Homo* species or unique to the fossil itself that indicate its distinctiveness from other *Homo* species. Ironically, incompleteness may also be the reason as to why these fossils were classified as *Homo* in the first place. Had these fossils been more complete they may have shown features that would have excluded them from *Homo*. Considering that at the time interval these fossils are found, differences in teeth or jaw fragments between *Homo* and other contemporary closely related lineages may not be evident. More complete fossils are necessary to arrive at unambiguous classifications.

Large brained hominids with limb bones and hand and foot bones which are very human-like appear in deposits, dated to around two million years ago, all along the east African rift. These fossils are best known from Olduvai in northern Tanzania and Lake Turkana in Kenya, but also may be found as far south as South Africa and as far north as Ethiopia.

A TWO LEG ADVANTAGE

It had not been an unusual year until the rains came. In the mountains to the north, it rained for three days straight. The river overflowed its banks, flooding many miles of land around its delta where it entered the lake. Out in the rain, a herd of lechwes waded in flooded grassland, nipping at the shoots peeking above the water line. A man, knee-high in water, was standing on two legs pulling plants out by the roots and tossing them onto land. Five other men sat immobile below a patch of bush-willow trees, watching and trying to avoid the rain. A troop of baboons watched the men at a safe distance. By late afternoon the sun came out, but it did not stop raining. The rain continued even with a full amber moon lighting the sky at night. The red sky that evening, however, was a sure sign that the storm would not last through the night.

The next morning the sun shone brightly, although isolated clouds with a flat grayish base still lurked in the sky. The lechwe herd had grown larger and could be seen at a distance feeding along the water's edge. A white rhinoceros and two crown cranes fed in the area where the man stood the previous day. The water had receded and the grasses claimed back some of the flooded land. Several men sat feeding on nightshade and wild lettuce leaves that were growing in partial shade within a stand of saplings.

The following day the water retreated fully, laying bare grassland and an expansive mudflat. The men ventured out onto the mudflat on two legs, as if they were wading in water. In one place the water-saturated mud gave way, and one of the men sank almost up to his knees. In the muddy soil, plants could be easily pulled out by the roots. The group wasted no time in doing so. They sat in thick growth muddied by the flood, feeding on grass roots. A marabou stork pecked away at the muddy ground a short distance away from them.

Curious to see what the marabou stork was busy with, one of the females in the group walked towards it, shooing it away. As she got closer, she noticed a fish lying on its side in a shallow puddle of water. She quickly grabbed for the fish, but it squirmed away from her, jumping back into the puddle. After repeated tries, she was finally able to get her hands on it and immediately brought it up to her mouth. Biting its back, she made sure it couldn't get away. While the fish was still alive she chewed off a piece. Her two sons who had been watching walked over, and the fish was passed from hand to hand until it was all eaten. The mother and her sons now looked for other fish. The youngest son had picked up a stick. Stunning the fish by beating them with the stick made it easier to catch them.

In the late afternoon, the baboon troop came to see what the men were eating. Venturing out onto the mudflat two baboons sank up to their chests and the whole group retreated. Keeping a safe distance, the long-legged marabou stork fed side by side with the men until sunset.

HOMO RUDOLFENSIS

Skull, Teeth, and Diet

Most of what is known of the *Homo rudolfensis* skull comes from one skull without a lower jaw bone (KNM-ER 1470) and a lower jaw bone (KNM-ER 1802), both from the same East Lake Turkana Kanopoi locality (area 131), Kenya. Other fossils assigned to *H. rudolfensis* are very incomplete and provide very little information. Although the KNM-ER 1470 skull appears reasonably complete, its surface is heavily abraded and preserves very little detail. The face is long and broad with well-separated eye sockets. A bar of bone stretches above both eye sockets, forming a single continuous brow ridge, which is thickest at the midline. The cheek bones are wide, from both side to side and top to bottom, attaching low on the upper jaw just behind the last premolar. The plate of bone holding the roots of the upper front teeth is set in the plane of the face. It is broad from top to bottom, providing wide separation between the mouth and the nose opening. Because the contact between the face and the skull is tenuous, it is not certain if the face and upper jaw protruded out as in *Paranthropus aethiopicus* or were tucked in and more vertically set as in *Paranthropus boisei*. The back of the skull and the skull base are similar in many respects to those of other *Paranthropus*. The braincase is relatively large, with a volume of 752 cc. Braincase narrowing behind the eye sockets is relatively mild. A much more fragmentary skull (KNM-ER 3732) yielded a slightly lower brain volume of 750 cc.

The KNM-ER 1470 skull has no teeth. Only the tooth sockets containing the sheared-off tooth roots are preserved. The front tooth row, including the canines, is set in a straight line parallel to the plane of the face and at nearly right angles to the right and left cheek tooth row, as is common in *Paranthropus*. The cheek tooth rows are short from front-to back and widely set apart, resulting in a very broad palate. The last molar appears to be set much higher than the cheek teeth in front, indicating that it may not have been fully erupted. Because all the sutures between the skull bones are not yet fused, KNM-ER 1470 was probably a young adult.

The lower jaw bone (KNM-ER 1802) more or less mirrors the shape of the KNM-ER 1470 skull palate. The front tooth row and canines are set nearly in a straight line at right angles to the cheek tooth rows. The molars and premolars are relatively wide from side to side. In many features, the premolars resemble those of *Paranthropus*.

There is precious little evidence bearing on the diet of *Homo rudolfensis*. Similarities to *Paranthropus* in what is present of the teeth and palate suggest a similar diet. *H. rudolfensis* would have eaten a very wide range of food items many of low nutritional value. Due to a short palate and a mouth that probably could not be opened very widely, food items would have been relatively small or broken into small pieces by hand. Leaves, seeds and roots, tubers and possibly pith would have been its main staples. It probably would have fed opportunistically on fruits, large insects, and small vertebrates, gorging on these whenever they were available in large quantities.

Skeleton, Gait, and Posture

Aside from skulls and lower jaws, there are no other known skeletal remains of *Homo rudolfensis*. Isolated limb bone finds from The Upper Burgi (the deposit in which H. rudolfensis is found), however, are plentiful, and there are some skeletons associated with teeth and skull fragments. One partial skeleton (KNM-ER 3735) has been assigned to *Homo habilis*, and the other (KNM-ER 1500) to *Paranthropus*, but in both cases these classifications are controversial and far from certain. Nevertheless, both of these individuals are relatively small and do not match in appearance or size some of the larger isolated bones. These include a pelvis (KNM-ER 3228), the large lower limb bones (KNM-ER 1472 and KNM-ER 1481) described with the KNM-ER 1470 skull, a thigh bone missing joint surfaces (KNM-ER 3728), and the upper part of a forearm bone (KNM-ER 3956).

The pelvis has many features in common with modern humans and contrasts strongly with that of *Australopithecus*. Principally its thigh bone and backbone joint surfaces are relatively large. The lower limbs, therefore, could have supported the entire weight of the body, without help from the upper limbs. The thigh bones are slightly stockier than average for humans, but are human-like in having 1) a well-formed carrying angle (*see Homo sapiens*), 2) a tear-shaped mid-shaft cross-section without front to back thigh bone shaft flattening, 3) a relatively high neck angle, and 4) a relatively large hip joint surface. The relative orientation of the knee joint to the hip joint (around the long thigh bone axis) was such that the kneecap faced front when the hip joint was extended. These thigh bone features are associated with supporting the body weight on a single lower limb and maintaining both thigh and lower leg movements in the plane of forward motion. The shin bones (KNM-ER 1481) indicate a human-like ankle joint, orienting the foot's plane of movement more or less in the same direction as that of the lower limb. All of the hip joint and lower limb features are associated with moving about on the ground on two legs as do humans.

Body weight and stature calculations based on thigh bone girth and length relationships in modern humans indicate the KNM-ER 1472 and KNM-ER 1481 individuals had a weight of 47 kg and 46 kg, and a stature of 149 cm and 147 cm. The KNM-ER 3728 thigh bone, which in all respects is similar to the other two, would have corresponded to a 45 kg, 145 cm tall individual.

Given that there are no other certain *Homo* species known from skull or tooth remains of a corresponding size in the Upper Burgi Member where the hip and thigh bones were found, these may very well belong to *Homo rudolfensis*. If this is the case it is likely, *H. rudolfensis* spent a considerable amount of time on two legs.

Fossil Sites and Possible Range

Homo rudolfensis fossils are found in the upper Burgi Member of the Koobi Fora Formation in east Lake Turkana, Kenya. The upper Burgi Member represents river delta and lake sediments of relatively slow moving water. Other fossils (KNM-ER-819, a partial lower jaw, and KNM-ER 1590, a partial juvenile skull), which have been claimed to belong to *H. rudolfensis* and would document this species in other members of the Koobi Fora Formation, are too incomplete to assign with confidence to *H. rudolfensis*. As with all other sedimentary members of the Koobi Fora Formation, volcanic ash layers within the upper Burgi Member allow radioisotope dating of fossil remains.

Age

All *Homo rudolfensis* fossils lie below the KBS (Koobi Fora Base Stratum) Tuff, which has a radioisotope date of 1.88 million years. The fossils are thus slightly older than this layer of volcanic ash. Overlying a volcanic ash layer with a radioisotope

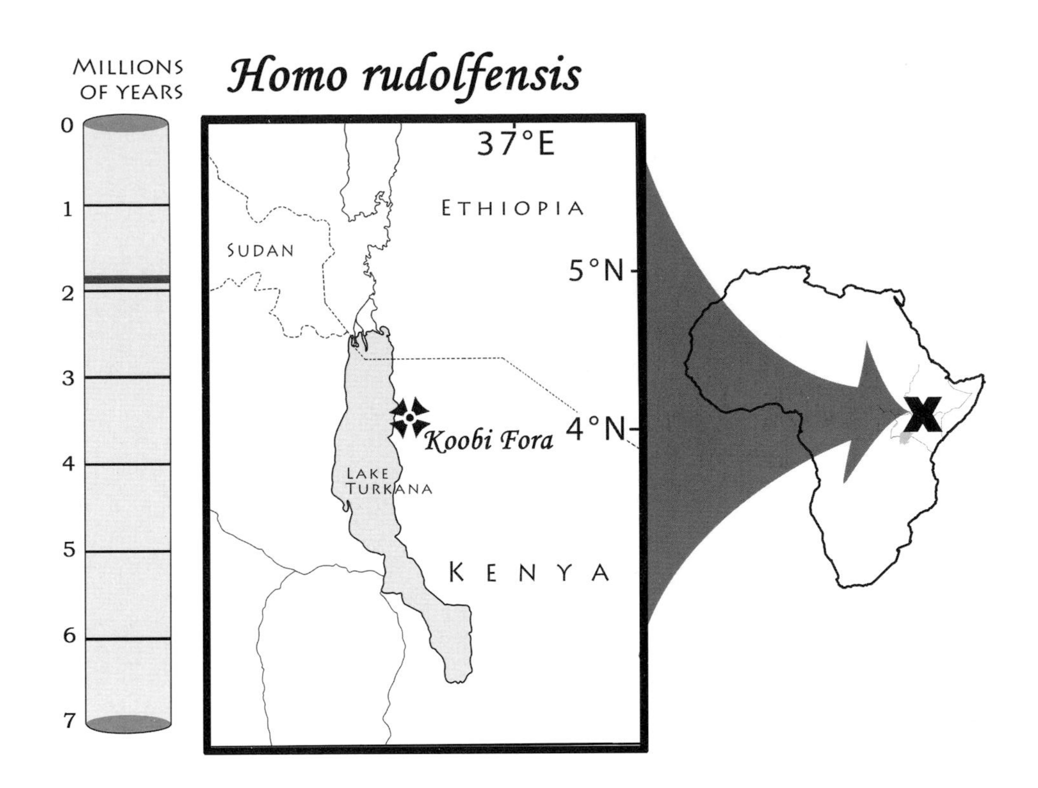

date of 1.9 million years (Lorenyang Tuff), all these fossils can thus be confidently dated to be between 1.88 and 1.9 million years old. All of the large limb bones from the upper Burgi which may possibly belong to *H. rudolfensis* are also from this time period. The hipbone (KNM-ER 3228), which was found below the Lorenyang Tuff, is slightly older, with an approximate geologic age of one million nine hundred and fifty thousand years.

Tools

There were no tools found in the *Homo rudolfensis* deposits. Choppers, hammerstones, and flakes, however, have been reported from the slightly younger KBS Member and from deposits predating the upper Burgi in west Lake Turkana. Although it is possible if not likely that H. *rudolfensis* would have made and used tools, there is no direct evidence to support that this was the case.

Animals and Habitats

As is the case for South African *Australopithecus*, many of the extinct genera found with *Homo rudolfensis* are monkeys and carnivores. These include a prehistoric baboon (*Parapapio*), two species of a ground-dwelling leaf-eating monkey (*Cercopithecoides*), two leaf-eating monkeys (*Rhinocolobus* and *Paracolobus*), a false saber-tooth (*Dinofelis*), and the scimitar cat (*Homotherium*). Extinct plant-eating genera include the three-toed horse (*Hipparion*), the short-necked giraffe (*Sivatherium*), three warthog species (*Metridiochoerus*), one bush pig species (*Notochoerus*), two buffalo species (*Pelorovis*), a prehistoric kob (*Menelikia*), and a giant hartebeest (*Megalotragus*). Modern animal genera include a gelada baboon, the black-backed jackal, the striped hyena, a laughing hyena, an African elephant, two Indian elephant species, two horse species, the white and the black rhinoceroses, two species closely related to the pygmy hippopotamus, a hippopotamus, four giraffe species, a kudu, two bushbuck species, three kob species, a lechwe, a roan, the oryx, a topi, hunter's antelope, two impala species, a springbok, four gazelle species, and a steenbok. A member of the reedbuck family resisted classification, as did three members of the gnu/hartebeest group that were recognized as distinct based on body size differences (i.e., small, medium, and large).

The variety of hippopotami and of kobs (including the lechwe) indicates a flooded grassland or lakeshore. The horses, warthogs, gazelles, gnus, hartebeests, oryx, and springbok all indicate open grassland. Overall, the animal assemblages suggest an abundance of open habitat with flooded grasslands and some woodland with heavy underbrush close to water.

Climate

The *Homo rudolfensis* sites are on the opposite side of Lake Turkana from Lodwar but close enough in distance to share more or less the same climate (*see Australopithecus anamensis*). Plants and animals in the *H. rudolfensis* deposits suggest the climate was more or less the same when *H. rudolfensis* inhabited the area as it is today.

Classification

Without any associated skeletal remains, a large braincase is the single most important feature justifying this fossil's classification in *Homo*. Superficial resemblances in its face to *Kenyanthropus* have suggested to some that it belongs in this genus and not in *Homo*. This, however, rests on the slim possibility that *Homo rudolfensis* has very non-human hip and limb bones and this does not seem likely. Certain association of the *H. rudolfensis* skulls to the large human-like skeletal bones found in the upper Burgi Member would clinch its position as a human ancestor and the earliest known member of the genus *Homo*. Considering the similarities between the *H. rudolfensis* and *Paranthropus* skulls, and the similarities between subsequent *Paranthropus* (*see Paranthropus boisei*) and *Homo* skeletons, *H. rudolfensis* would appear to be very close to the common ancestor of both *Homo* and *Paranthropus*. Regardless of its incompleteness, there is little doubt that *H. rudolfensis* is a member of the human lineage, post–African ape divergence.

Historical Notes

Bernard Ngeneo is credited with finding the first remains of *Homo rudolfensis*, the KNM-ER 1470 skull in 1972. In an April 1973 *Nature* issue, Richard Leakey reported on the skull together with the lower limb bones (some of which were complete) of three other individuals found in the same Lake Turkana location, asserting them all to be securely dated at 2.6 million years ago. Because Leakey believed the skull was too large-brained to be *Australopithecus* and was much too old and too large-brained to attribute to *Homo habilis*, he assigned it to *Homo*, but shied away from giving it a species name. In a 1973 *Journal of Human Evolution* paper, Alan Walker referred the thigh bones found in this same location to *Australopithecus*. Owen Lovejoy, in various publications, would later cite these thigh bones to argue that *Australopithecus* had human-like two-legged postures and movements. In a February 2003 *Nature*, R.J. Blumenschine and colleagues reported that a maxilla with a complete dentition (OH 65) found in the upper part of Bed I Upper Member Olduvai was similar to *H. rudolfensis*, indicating this species was synonymous with *H. habilis*.

In his 1986 book *Origin of the Human Race* (Progress Publishers Moscow), V.P. Alexeev gave the KNM-ER 1470 skull its species name, "*rudolfensis*," commemorating Lake Rudolf where the fossil was found. Unbeknownst to Alexeev, Lake Rudolf had been renamed Lake Turkana many years prior to his choice for a species name.

Thoughts of tomorrow underlie the intelligent gaze of *Homo rudolfensis*.

THE LUCK OF THE PYGMY

The grey blue water was lined by the pink hues of hundreds of flamingos feeding at the lake's shallow end along its shore. On the water saturated mudflat a young pygmy searched for nests among the salt cedars, hoping to find some eggs. When his search produced no results, he turned his attention towards the birds. Four birds were closest to him, standing apart from the rest. Familiar with flamingos, he knew how much running room they need to take off. He was sure he could chase one down, if he got close enough. Hiding behind the lakeshore growth, he attempted step-by-step to creep up on them. Acting in unison, the four birds maintained a safe distance, taking a step back for every step the pygmy took forward. Aware that he wasn't making any progress, he lost his patience and made a dash for them. Flapping their wings, the flamingos ran first in and then on the water, before finally picking up their legs and achieving free flight. The flight of the four alarmed the other flamingos along the shore, triggering a wave of flapping wings. Evenly spaced, hundreds of flamingos flew above the lake waters slowly gaining in elevation. As they turned away from the morning sun, their wings emanated an orange-pink light that flooded the sky.

Not discouraged by failure, the pygmy walked along the shore looking for other opportunities. Not finding any, he thought about all the food in the highland forest. It was early in the morning and the forest was close enough to make it there and back in a single day.

Walking away from the lake onto the scrubland, he came upon seven giraffes. All but a cow and her recently born calf took off running. The calf appeared to be lame and unable to run. As he came closer, the cow shielded the calf behind her. Yelling and waving a large leaf-covered branch, the pygmy was able to spook the cow, which reluctantly left its calf behind. Making a dash for the calf before the cow doubled back, the pygmy suddenly stopped in mid-stride. Thorns from acacia scrub covering the ground had penetrated the soles of his feet. He sat on the ground, legs crossed, pulling thorns out. With every thorn that came out a drop of blood was shed. All the while the calf bleated for its mother. With sore feet the pygmy tried to walk in the direction of the calf, but there were so many thorns on the ground he was unable to. He was forced to walk out of the scrub that surrounded the calf by the same route he walked in. Circling around the calf, he tried to find a thorn-free route. By now the cow had returned and once again had placed herself between him and the calf. With branch in hand, he tried a new route to approach the calf, but once again as he came near, the thorns grew denser. Feeling the pain in his feet, he gave up the hunt, walking further out into the scrubland.

In the distance, the pygmy saw two of his brothers. He called out and in time caught up with them. The three now moved towards a flock of vultures circling in the sky close to the horizon. More than an hour passed before they reached the area over which the vultures had been flying. What appeared from a distance to be a mound of soil was the back of a bull elephant lying on its side. The vultures were now patiently waiting, perched in the leaf-less branches of a dead tree. As the brothers approached, they could see the elephant was still alive and forcefully breathing. A broken tusk was impaled in its skull just behind the eye-socket. As evidenced by the thick clear secretion that had dripped from its temple and streaked its jaw and forelimb, the elephant was in musth. It had probably sustained its life-ending injury fighting another musth bull. The wound and the pool of blood that collected on the ground below the head had attracted a swarm of flies. Their buzzing was deafening and masked all other sounds. A lone jackal sniffed at the puddle of blood collecting at the elephant's underside.

Excited at what they saw, the three brothers hunted for sticks and stones and returned to beat the elephant to death. Banging rocks together, the oldest brother produced sharp flakes to cut the elephant's skin and get at the meat below. The youngest brother pulled back the cut skin. The middle brother caught and ate the parasites on the elephant's skin. By late afternoon more than thirty pygmies had congregated around the elephant corpse to feed. In the distance, a hyena was circling, taking a roundabout way to reach the elephant. The pygmies would spend the night with the elephant corpse, eating and chasing away with sticks and stones hyenas, jackals, and other competitors.

HOMO HABILIS

Skull, Teeth, and Diet

Homo habilis skull features have been defined relative to *Australopithecus* and to the closer approximation these features show to the human condition. In general, the *H. habilis* face is slightly built when compared to that of *Australopithecus*, but the eyes are nonetheless widely set apart. *Homo habilis* has a continuous bony brow ridge above both eye sockets, although it is not as well formed as in *Homo erectus*. Behind the eye sockets, the *H. habilis* braincase does not narrow as much as in *Australopithecus* or *Paranthropus*, presenting a larger braincase volume. Based on four individuals, *H. habilis* exhibits a mean braincase volume of 641 cc with a range of 590 cc to 687 cc.

Along the midline at the base of the *Homo habilis* nose opening, there is a strong spine for the attachment of the nose septum as is seen in other *Homo* species. In addition, the plate of bone holding the roots of the front teeth is vertically set and widely separates the mouth from the nose opening. Compared to *Australopithecus*, the molars and premolars are absolutely smaller and proportionately narrower, with premolars that are narrower from side to side than molars. Relative to molar row length, the front tooth row is much wider than in *Australopithecus*.

Neither micro-wear nor carbon isotope analyses of *Homo habilis* teeth aimed at understanding their diet have been undertaken. Overall, the diet of *H. habilis* was probably not much different that of *Australopithecus*. The smaller molars, given what appears to be a comparable body size to *Australopithecus*, suggest foods that would not need as much chewing, because either they could be easily broken up by the teeth into smaller parts or they had a higher nutritional content and thus smaller amounts were needed. Associated tools and cut marks on bones found in the *H. habilis* deposits suggest that such higher quality food could have consisted of meat from large vertebrates.

Skeleton, Gait, and Posture

Most of the presumed *Homo habilis* skeletal bones have not been found associated with skulls or teeth. With both *H. habilis* and *Paranthropus boisei* present in the same deposits, it is not known which fossils correspond to which species. Given that some Olduvai fossils show a chimpanzee-like anatomy and others a human-like anatomy, there is considerable uncertainty as to whether any one species possesses a mosaic structure of chimpanzee-like and human-like features, or whether all the human-like remains correspond to *Homo* and the chimpanzee-like remains to *Paranthropus*.

The OH 7 hand is an exception in that it was found associated with teeth and skull bones. As in humans, the OH 7 hand is wide, with a relatively large thumb and broad fingertips. It differs from humans, however, in having relatively large fingers with chimpanzee-like curvatures and a thumb, which, despite its human-like size, has a more rotated great ape orientation relative to the other fingers. The OH 8 foot is very similar to that of humans in showing limited mobility at most of its joints and a big toe, which is held more or less parallel to the other toes. Although the foot lacks a human-like arch and its owner would have had a flat-footed stance, the foot had lost its grasping ability, with its versatility sacrificed for more efficient ground movement. The lower leg bones (OH 35) have been described as human-like in most respects. Length estimates provided in this original description, however, were overestimates. Mid-shaft-diameter to length ratio is more like that of living great apes than humans. The orientation of the ankle joint was such that the feet would have been strongly pigeon-toed when the kneecap faced forward. This foot posture is in marked contrast to the normal human condition in which the foot has some degree of toe-out. Modern humans with feet as pigeon-toed as the Oh 35 individual are unable to maintain their balance while walking. There is no conclusive evidence that *Homo habilis* customarilly moved around on two legs, even if one accepts that all the bones assigned to *H. habilis* belong to this species. The different ankle joint orientation also suggests that *Homo habilis* would have walked differently than modern humans.

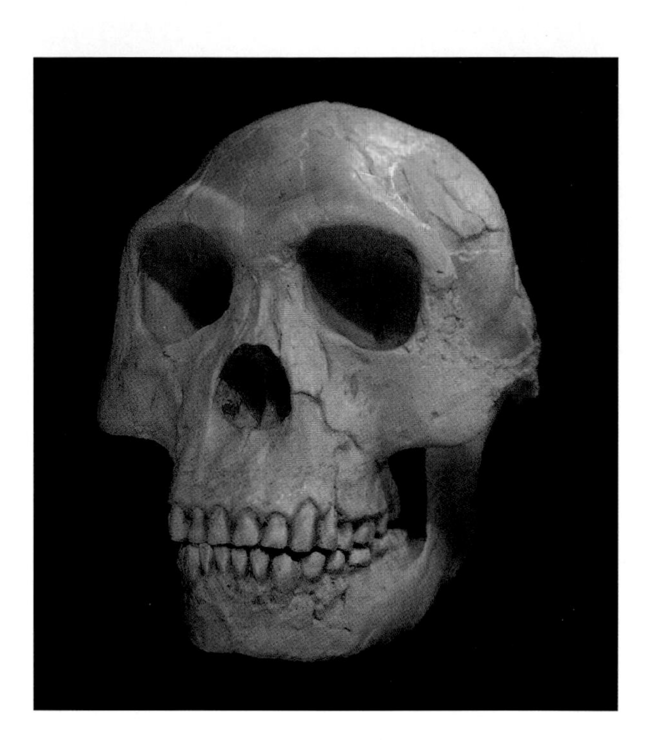

Composite reconstruction of the skull of *Homo habilis*, using parts from KNM-ER 1813, OH 24, and Sts 19 skulls and OH 13 jaw and teeth.

Mid-shaft diameters of the OH 62 thigh bone and arm bone have more or less the same value, suggesting that *Homo habilis* upper limbs may have supported as much of the body weight as its lower limbs. These mid-shaft diameters also indicate it was slightly lighter in build than A.L. 288-1 (Lucy). Claims of upper to lower limb length ratios, which in one analysis are more human-like, and in another analysis more chimpanzee-like than those of *Australopithecus*, must be taken with a grain of salt. These proportions cannot be accurately reconstructed with most of the bone shaft missing.

Fossil Sites and Possible Range

Homo habilis fossils come from the upper member of Bed I and lower and middle members of Bed II at Olduvai Gorge, Tanzania. These deposits represent sediments from a small lake (200 km^2), which formed in the Olduvai region of Tanzania about two million years ago and remained there for at least four hundred thousand years or more. With its water source coming from streams in the surrounding volcanic mountains and no outlet to other bodies of water, the lake had a water level that fluctuated considerably with rainfall and at times had extremely alkaline and saline conditions. During Bed II deposition, the lake shrank to approximately a third of its former size, leaving a flooded-grassland in parts of its former extent.

Throughout the deposits volcanic ash layers with interlayer clay sediments provide means for absolute radioisotope dating of fossils. Movements of Earth's crust, the result of the African continent rifting apart, have produced various faults in the sediments. These have made it more challenging to correlate layers exposed at various localities and correctly date the deposits. With subsequent drying of the lake, the actions of the Olduvai River passing through carved out a gorge, exposing the sediments.

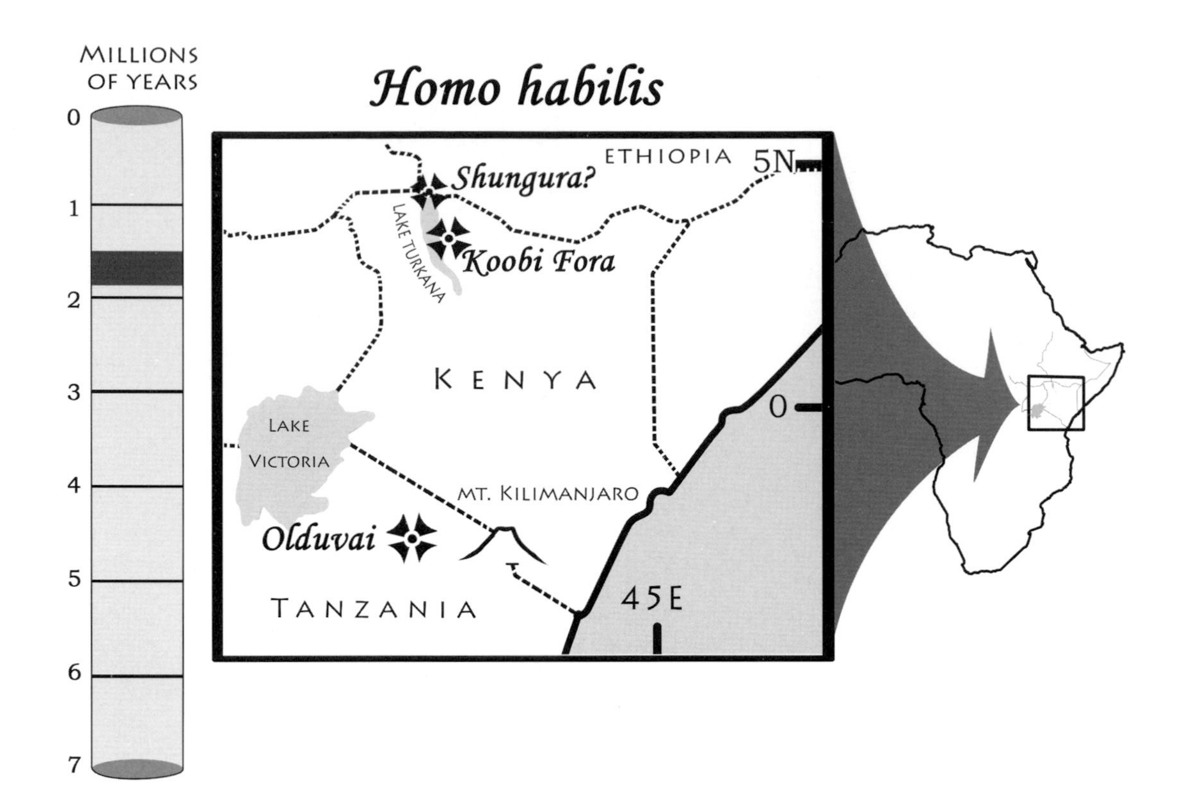

Homo habilis has also been reported in the upper Burgi Member and KBS Member "channel complex" of the Koobi Fora Formation east of Lake Turkana, Kenya, at 1.88 million to 1.9 million and 1.85 million years ago. *H. habilis* is also claimed to exist in the upper G Member of the Shungura Formation, Omo River basin, Ethiopia, at around one million nine hundred thousand years ago. With the exception of the KNM-ER 1805 and KNM-ER 1813 skulls, all these fossils are too incomplete to assign with any certainty to *H. habilis*. Both KNM-ER 1813 and KNM-ER 1805 have also been claimed to belong to *Homo ergaster*. Because the features distinguishing *H. ergaster* from *H. habilis* have not yet been fully defined, it is uncertain to which species these two skulls should be referred.

Age

An ash layer (Tuff I^{B}), underlying all *Homo habilis* fossils and overlying the basalt base of upper Member I, yielded radioisotope dates ranging from 1.75 million years to 2 million years ago (later revised to 1.83 million years). Ash layers (Tuff I^{A}) laid down below the basalt base of Bed I upper member with the appearance of the lake and, at the very top of Bed I (Tuff I^{f}), yielded radioisotope dates of 1.96 million years and 1.75 million years ago. The ash layer between Bed II lower and middle member (Tuff II^{A}) yielded a date of 1.7 million years, but because of contamination from other ash this date is not very certain. Paleomagnetic analyses, therefore, were used to verify and further fill in the dates. During deposition of upper member Bed I, the sediments record a normal orientation of Earth's magnetic pole. The beginning of this normal period is known to occur at 1.94 million years ago in deposits throughout the world. At Olduvai its beginning is found 6 m below Tuff I^{A}. A half meter above the Bed II base, the deposits show a reversal in Earth's magnetic pole. This reversal is known to occur 1.79 million years ago. All of upper member Bed I deposition, therefore, occurred between 1.94 million and 1.79 million years ago. The Bed II lower member has an age between 1.79 million and 1.7 million years ago. The end of the Bed II middle member probably has an age of 1.53 million years. All the *H. habilis* fossils, therefore, date to between 1.83 million years and

A brilliant African sky offers a visual wonder for a curious female *Homo habilis*.

1.53 million years ago. Possible *H. habilis* fossils from the Omo Basin and Lake Turkana are thus older than those from Olduvai.

Tools

Throughout its lifespan at Olduvai, *Homo habilis* is associated with tools. In fact, its tool association provides an important part of its definition as a species. Tools found in Bed I upper member are of the Oldowan tradition, consisting of hammerstones and simple flakes. Lower Bed II shows more advanced Oldowan tools, referred to as Developed Oldowan, in which stones are worked on two sides producing bifaces. Animals bones

found in the *H. habilis* deposits show cut marks indicating meat was stripped from the bone. What looked to be cut marks made on bone overlying carnivore tooth marks were interpreted to mean *H. habilis* was scavenging dead animals or stealing animal kills from large carnivores. Recently, this conjecture was shown to be unsupported, since supposed cut marks and some tooth marks were most likely the result of bone weathering.

Animals and Habitats

Animals found in the Olduvai Bed I and lower Bed II with *Homo habilis* are modern in appearance.

Tired, starving, and weary, this male *Homo habilis* has just caught the scent of a carcass in a nearby field.

Modern genera include a hedgehog, five shrews, a short-nosed and a long-eared elephant shrew, a twilight and a pipistrelle bat, a baboon, a bush baby, two species of gelada baboon, a ground squirrel, a spring hare, a porcupine, a naked mole rat, a pygmy gerbil, a naked-soled gerbil, a swamp rat, seven rat/mice species, a bat-eared fox, a laughing hyena, a lion, a dog, a black-backed jackal, a river otter, an uncertain number of genets, an African mongoose, a cusimanse, a white-tailed mongoose, two Indian elephants, a white rhinoceros, a horse, a hippopotamus, a bush pig, two warthogs, a giraffe, an okapi, a nyala, a giant roan antelope, a Hunter's antelope, a topi, a wildebeest, and a gazelle. Most of the animals either represent extinct species or resisted species identification. The black-backed jackal, however, is the same species as the animal that is still found in the Olduvai area today. There are, however, some extinct genera found with *H. habilis*. These include a prehistoric cousin of horses and tapirs (*Ancylotherium,* a Chalicothere), a pig (*Kolpochoerus*), a hartebeest relative (*Parmularius*), a prehistoric elephant with lower jaw tusks (*Deinotherium*), a prehistoric horse (*Stylohipparion*), an unidentified saber-tooth cat, and *Paranthropus boisei*.

Some of the Olduvai animals are typically found close to water or in flooded grasslands (hippopotamus, white rhinoceros, swamp rat, the white tailed mongoose, and the river otter) and reflect a lakeshore habitat. The other animals indicate savannah grasslands with patches of open woodland and denser bush close to streams and lakeshores. Notably, the majority of small mammal genera still exist in the area today, reflecting more or less similar habitat and climatic conditions.

Climate

The closest weather stations collecting long-term records are 200 km away from Olduvai Gorge, so that the following is based on generalized averages. Mean annual precipitation is 500 to 750 mm. March-April receives about a third of the annual rainfall (167 mm to 250 mm), with June-September receiving about 70 mm. Average mean monthly temperatures vary relatively little throughout the year. The coldest months, June and July, record a monthly average of 16.7° C (62° F) and the warmest months, December-March, record an average of 21.7° C (71° F). Temperatures are always above freezing, with a record low of 7° C (45° F), occurring during August. There is no reason to expect that rainfall or temperatures would have been much different when *Homo habilis* lived.

Classification

Unfortunately, Wilfrid Le Gros Clark's anti–*Homo habilis* influence has played havoc with African *Homo* classification. Many paleoanthropologists, heeding Le Gros Clark's wishes (*see* below, Historical Notes), have preferred to refer all *Homo* fossils to *Homo erectus* or to new species (e.g., *Homo ergaster, Homo rudolfensis*, etc.), without entertaining the possibility they could be *H. habilis*. Assigning African fossils to *H. erectus*, a castaway species on an island 10,000 km away, which is represented by very incomplete remains and may be as much as a million years younger than many African *Homo* species, appears to be carrying Le Gros Clark's wishes a bit too far. In spite of such concerted efforts, *H. habilis* has refused to go away.

Considering that its inclusion in *Homo* is also based on hand and foot bones, it is one of the most accurately described species of early humans, much more so than *Homo erectus*. *Homo habilis* serves as an example that those planning to name new fossil species should emulate, especially when looking to do so based only on skull remains.

Homo habilis, however, is not devoid of problems. Most of the fossil skulls upon which this species is based are those of females and/or immature individuals, challenging paleoanthropologists to imagine how the appearance of females and immature individuals translates into that of a large male. More importantly, its validity as a species rests on the likelihood that the OH 8 foot bones (*see* above, Skeleton, Gait, and Posture) do indeed belong to *H. habilis* and not *to Paranthropus boisei*. Coming from the same beds as all the other *H. habilis* remains, the recently described OH 65 palate is best referred to this species until analyses show otherwise.

Historical Notes

The first remains of *Homo habilis* to come from Olduvai Gorge, a lower jaw fragment with a single molar and two broken teeth, catalogued OH 4, were found by H. Mukuri in June of 1959. In a November 1962 issue of *Nature*, John Napier described the OH 7 hand bones as an unknown hominid species lacking full development of the human-like precision grip. In a March 1964 *Nature* issue, Napier and Michael Day described the OH 8 foot as "hominine," meeting the physical requirements of a human-like two-legged stance, but without having yet achieved the human-like stride. In an accompanying paper, P. R. Davis described associated lower leg bones (OH 35) as belonging to a hominid that routinely moved about on two legs, but with a knee joint that may not have been as well adapted to such movement. In a 1964 April issue of *Nature*, Louis Leakey, Phillip Tobias, and Napier described the new species *H. habilis,* designating OH 7, nicknamed "Johnny's Child" (a lower jaw without erupted wisdom teeth and associated skull fragments, hand bones, and an upper molar), as the holotype specimen. Relying in part on previously described and analyzed material, they included within this *H. habilis*: 1) OH 4, a lower jaw fragment; 2) OH 6, a lower premolar, upper molar, and associated skull fragments; 3) OH 13, an incomplete skull nicknamed "Cinderella"; 4) the OH 8 hand and foot, and associated collar bone fragment; 5) OH 14, skull fragment preserving a jaw joint; and 6) OH 16, another partial skull, nicknamed "George." They also revised the definition of *Homo* to better differentiate it, and the new fossils they included in it, from *Australopithecus*. At the suggestion of Raymond Dart, "*habilis*" (which in Latin means "able, handy, mentally skilful, vigorous") was chosen as the species name to emphasize this hominid's mental capacity and toolmaking skills.

In a May 1964 *Nature*, Kenneth Oakley and Bernard Campbell cautioned the describers of *Homo habilis* that before creating a new species careful comparative investigations with closely related species must be undertaken. Shortly thereafter, Wilfrid Le Gros Clark wrote in a July 1964 issue of *Discovery* that he hoped *H. habilis* "will disappear as rapidly as he came." Many paleoanthropologists took this task to heart and referred to *H. habilis* as *Australopithecus habilis* well after Le Gros Clark died. In 1968, an additional skull, OH 24, nicknamed "Twiggy," was found in Olduvai lower Bed I upper member. In a May 1987 issue of *Nature*, Donald Johanson and colleagues reported finding a very incomplete *H. habilis* skeleton (OH 62) in the lower Bed I upper member deposits, which they claimed had more chimpanzee-like limb length proportions than Lucy (A.L. 288-1). In 1991, Phillip Tobias published an exhaustive description of the four *H. habilis* skulls and other claimed *H. habilis* material in the IVa and IVb Olduvai volumes (Cambridge University Press).

Homo habilis fossils are kept at both the National Museums of Kenya, Nairobi, and the National Museum of Tanzania, Dar es Salaam.

IN THE SHADOW OF MAN

Over the last three decades the lake had been drying up and shrinking in size. Although it seemed that the number of animals using it had increased, actually there were about the same number, but crowded into a smaller and smaller space. During the dry season, animals from the surrounding grassland came to the lake to drink and feed on the vegetation the water promoted. Without rain, the grassland began to dry up and produced very little food to support the animals. At this time, the high density of plant-eating animals the lake attracted also attracted predators. The dry season for most animals was a time of food shortages, but for predators the high density of animals around the lake made it a time of food surplus.

Except for large predators, therefore, the lake was a dangerous place. Most animals found safety in numbers. Animals of different species often drank or fed together in large parties. They gambled on the hope that another animal would be the one to satisfy the predator's hunger. Danger, however, was not the exclusive provenance of predators. The retreating shore line crowded male hippos, staking out their territory in the lake shallows, into increasingly smaller areas. They became more aggressive, attacking any animals infringing on their space, perceiving them to be a threat.

A group of apemen, a gelada baboon troop, and two humans fed together on herbs and grasses along the shore, keeping a safe distance from the ramp-like hippo trails descending into the lake. With many of the grasses mature and turned to seed, the apemen concentrated on these, using their front teeth to strip the seeds from the tall tufts. Willing to brave their proximity to the hippo trail, the humans fed on the rootstocks of a small patch of sedges exposed by the receding shoreline. When an apeman came to feed next to them, the two humans became visibly uneasy, increasing their eye movements and averting a fixed gaze.

By late afternoon the apemen, the baboons, and the humans had moved away from the lakeshore and towards a strip of trees growing along a stream. A wildebeest and a spring hare now fed next to them. With the approaching evening the apemen, the two humans, and the gelada baboons separated, each group finding a place to sleep among the spaces the rocks and undergrowth under the trees afforded. A recently weaned youngster asserting its independence from its mother nested alone in a tree above the ground nests where the adults slept.

At sunup the apemen and the gelada baboons were together again feeding on herbs and grasses. The humans were nowhere to be seen. Around late morning the querulous calls of bulbuls feeding nearby alerted the apemen to the availability of fruit. Following the calls, they arrived at a half a dozen Natal plum trees ripe with fruit. Shortly thereafter, the shaking of branches and the rustling of undergrowth was followed by shrill screams. The two humans who the day before had been feeding with the apemen ran out of the grove onto the grassland, one of them carrying the body of a young apeman. Hiding behind some shrubs along the lakeshore, the two humans butchered the youngster with their stone tools. Eating part of it, they carried the rest back to camp.

Known to attract animals that eat fruit and the predators hunting them, fruiting trees further compounded the lake's danger. Living throughout most of their range with an erstwhile vegetarian turned into a stone-wielding predator, the apemen now found themselves in a much more dangerous place.

PARANTHROPUS BOISEI

Skull, Teeth, and Diet

The *Paranthropus boisei* skull is large, with a long face and a massive lower jaw. Despite their large size, the upper and lower jaws are tucked in below the face, which is almost vertically set. The cheek bones are very wide from side to side, and the eyes are widely separated by a very broad nose bridge. Owing to its large skull size, the braincase is also large relative to *Australopithecus*. Mean braincase volume based on six skulls is 508 cc, with a range from 475 cc to 545 cc.

Assigned to the same genus as *Paranthropus robustus* and *Paranthropus aethiopicus*, *Paranthropus boisei* shares many of the same skull and tooth features. The skull base is short from front to back and faces directly downward. Neck muscle attachments are well marked by bony crests and positioned low on the skull. On top of the braincase along its midline there is also a bony crest for the attachment of chewing muscles.

The cheek teeth are large with flat wear, and the premolars and molars approximate each other in both shape and size. The first milk molar is wide from side to side and has a large crushing surface. The front teeth and canines are relatively and absolutely small. Moreover, the canines 1) have flat wearing tips with a plane of wear at the same level as that of the other teeth, 2) are not associated to tooth row gaps or honing lower premolars, and 3) appear not to show differences between males and females.

Relative to *Paranthropus robustus*, *Paranthropus boisei* shows more extreme development of those features shared by the two that distinguishes them from *Australopithecus*. For instance, its molars are larger than those of *P. robustus* and its front teeth and canines relatively smaller. Its face and the attachment areas for most of its neck and chewing muscles also appear to be much larger. As such, *P. boisei* has often been referred to as hyper-robust. As seen in the excellently preserved OH 5 skull, *P. boisei* differs from *P. robustus* and resembles chimpanzees in having a sinus (air space) in the bone separating the roof of the mouth from the nose opening. It further differs from *P. robustus* in having canines and front teeth all with the same forward set, forming a straight line from side to side (canine to canine) nearly at right angles to the cheek tooth rows.

The diet of *Paranthropus boisei* has not been well studied. Given tooth and skull similarities to those of *Paranthropus robustus*, a similar diet is expected. Exaggerated development of distinguishing *Paranthropus* features, however, reflects a greater emphasis on eating small abrasive foods. Seeds, roots, and tubers would have probably been emphasized even more than in *P. robustus*. Considering *P. boisei* is found in lakeshore deposits surrounded by relatively open habitats with long dry seasons, fruits would not have been available year round. Nevertheless, fruits, leaves, insects, and small vertebrates would probably have been eaten when available. Whether or not *P. boisei* had a taste for meat and is responsible for the cut marks found on the bones of the large mammals with which it lived is a possibility that needs further investigation.

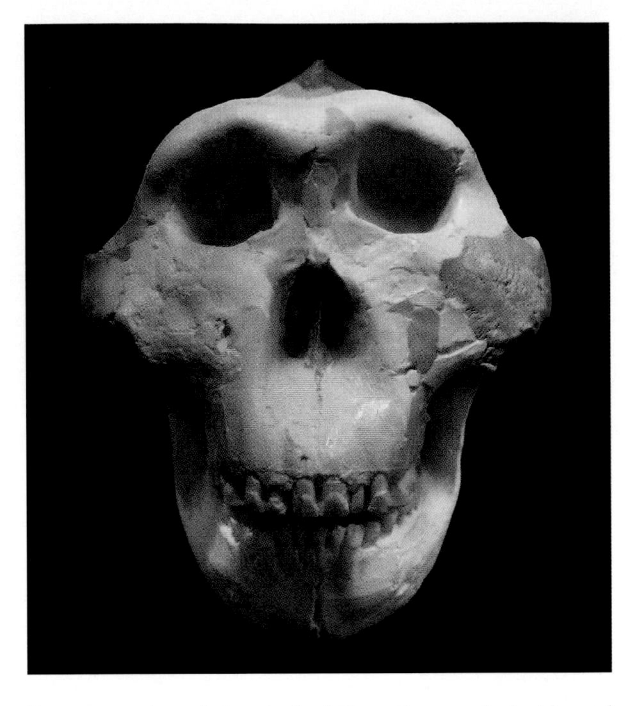

Reconstruction of the skull of *Paranthropus boisei* based on the famous "Zinj" skull (OH 5) found by Mary Leakey in Olduvai Gorge, with a modified Pening 1 mandible (jaw).

Skeleton, Gait, and Posture

In spite of being represented by some very complete skulls (OH 5, KGA 10-525, and KNM-ER 406), the skeleton of *Paranthropus boisei* is very poorly known. A complete ankle (Omo 323-76-898) and thumb bone (Omo 323-76-897) from the single associated skeleton that has been identified as *P. boisei* both appear to be similar to those of *Homo*. Confusion with *Homo* fossils due to their occurrence in the same deposits is probably the reason that so few *P. boisei* skeletal bones are known. For instance, the ankle bone (Omo 323-76-898) was recently reanalyzed and assigned to *Homo* by paleoanthropologists who were not aware of its association with the *P. boisei* skeleton. The OH 35 (originally OH 22) lower leg bones, found in the proximity of the OH 5 skull, were later associated with the OH 8 foot and assigned to *Homo* based mainly on their similarity to modern humans. As has been noted, many of the skeletal remains from Olduvai Gorge assigned to *Homo habilis* may actually belong to *P. boisei* (*see Homo habilis*).

An upper thigh bone (OH 20) from Olduvai Gorge is similar to the two Swartkrans proximal thigh bones (SK 82 and SK 97), suggesting it may belong to *Paranthropus boisei*. The *Paranthropus boisei* thigh bone, thus, may be distinguished from those of *Homo* by a neck set at a low angle to the shaft, a shaft that is flattened from front to back, and possibly a small hip joint surface. Many isolated thigh bones from Koobi Fora, therefore, can be assigned to *P. boisei*. Unfortunately, both shaft flattening and low-neck angles are also seen in the thigh bones of a supposed *Homo* skeleton from Koobi Fora (KNM-ER 1808) and of *Homo pekinensis* from Beijing, hopelessly confounding the situation.

Although similarities *Paranthropus boisei* shares with *Homo* can be used to argue it routinely moved around on two legs, the human hand, is dripping with adaptations for weight support and propulsion, and lower limb mechanical requisites for four-legged movement do not differ much from those for two-legged movement. Given the absence of skeletal bones that can be confidently assigned to *P. boisei*, it's movements and behaviors remain unknown.

Fossil Sites and Possible Range

Paranthropus boisei is known from upper member Bed I and possibly all of Bed II in Olduvai Gorge Tanzania; upper Burgi, Koobi Fora, and Okote Members of the Koobi Fora Formation in east Lake Turkana, and the Kaitoo Member of the Nachukui Formation in west Lake Turkana, Kenya; Member G and possibly H of the Shungura Formation, along the Omo River, and interval 4 sediments at Konso, in the southwestern end of the main Ethiopian rift 180 km northeast of Lake Turkana, Ethiopia. *Paranthropus boisei* has also been found in deposits at Chesowanja, east of Lake Baringo, Kenya, in Peninj west of Lake Natron, Tanzania, and in the Chiwondo beds of the Malema Valley west of Lake Malawi, Malawi (*see Homo habilis* and *Paranthropus aethiopicus*). These sites span elevation from 375 m to 1500 m above sea level.

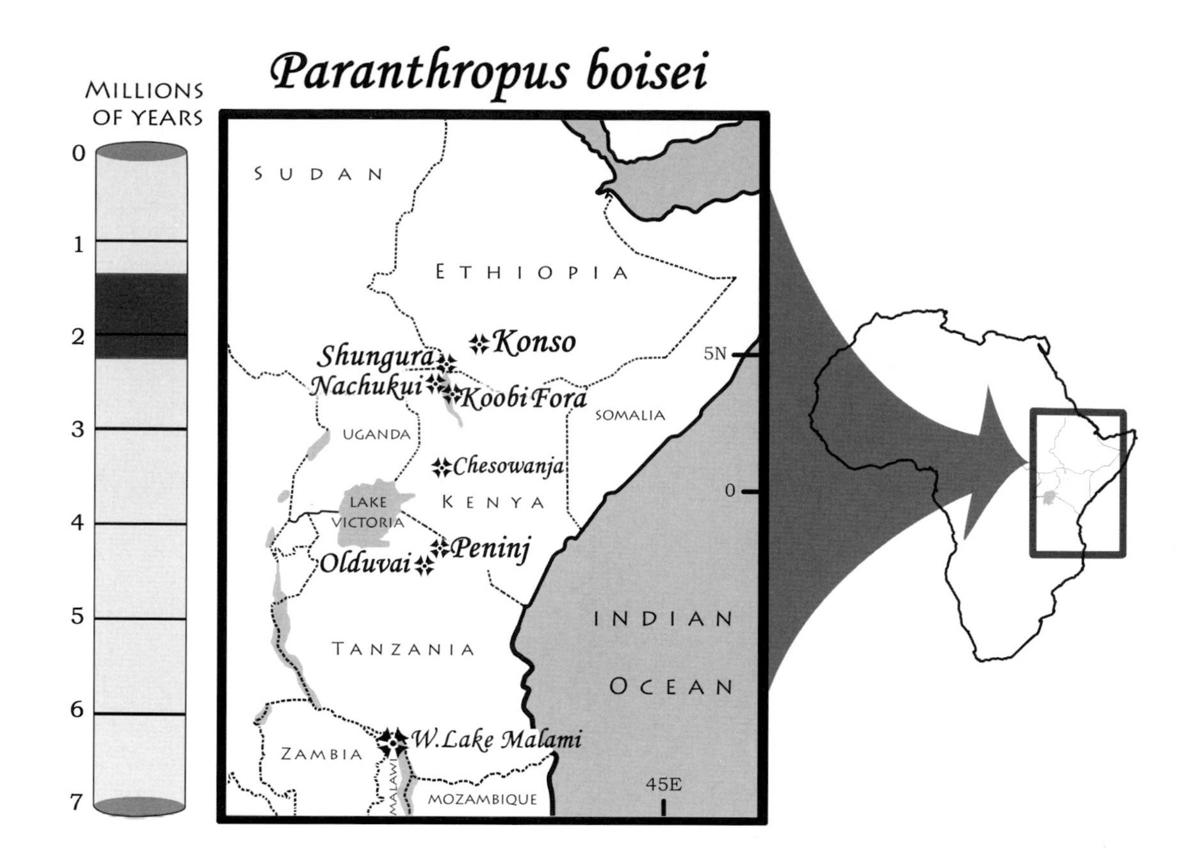

Age

Many of the *Paranthropus boisei* deposits have been dated by radioisotope and/or paleomagnetic means, and thus have accurate dates. At Koobi Fora, its remains span from 1.49 million to 1.9 million years ago. At Konso, they are dated to between 1.41 million and 1.43 million years ago. The associated *P. boisei* skeleton at locality 323 in Member G of the Shungura Formation, Ethiopia, has a date of 2.1 million to 2.3 million years. The Kaitoo Member of the Nachukui Formation, Lake Turkana, is bracketed by ash layers that indicate it was deposited between 1.8 million and 1.49 million years ago. By assuming a constant deposition rate (*see Australopithecus anamensis* Age) and given the depth in the deposit in which the *P. boisei* KNM-WT 17400 skull was found, its age has been calculated at 1.77 million years. As a surface find without associated sediments, the Peninj lower jaw cannot have a certain date, but Peninj fossil bearing strata have a radioisotope and paleomagnetic date between 1.4 million and 1.6 million years ago. The lower jaw is most likely to have come from these sediments and thus has a corresponding age. The Chesowanja skull was found slightly below a deposit with a radioisotope age of 1.42 million years, indicating it has somewhat slightly older age.

As noted (*see Homo habilis,* Age), the 1.83 million year age for the base of upper member Bed I at Olduvai Gorge is corroborated by both paleomagnetic and radioisotope analyses and is very accurate. The end of Bed II, however, presents problems both for dating and for confirming *Paranthropus boisei* occurrence. The only *P. boisei* fossils claimed to come from Bed II upper member are a milk canine and milk molar (OH3). Unfortunately, these fossils are too incomplete to be certain they belong to *P. boisei*, and their assignment as such has been controversial. If these fossils do not belong to *P. boisei*, then the upper limit for *P. boisei* occurrence is the lower member of Bed II, with a minimum date of 1.7 million years. A recent radioisotope analysis of the ash layer just above Bed II (Tuffs III-1) produced a date between 1.25 million and 1.47 million years ago. Considering that the end of Bed II lies below this volcanic ash and is somewhat older, an age of one million four hundred thousand years is a reasonable approximation for the end of Bed II. Assuming the OH 3 teeth are those of a robust, *P. boisei*'s most recent occurrence at Olduvai is 1.4 million years ago.

The Chiwondo beds at Malema Valley are the only *Paranthropus boisei* deposits that have not been absolutely dated. A date of 1.5 million years for the *P. boisei* fossils, based on comparison of the fossil animals found in the Malema locality with those of other deposits with known dates, is in accordance with absolute dates for *P. boisei*. In summary, *P. boisei* fossils span a time interval between a maximum of 2.3 million years and a minimum of 1.4 million years ago.

Tools

Simple stone tools consisting of choppers and flakes have been found in nearly every *Paranthropus boisei* fossil deposit. Because *Homo* is also found in these deposits, stone tools have been commonly attributed to *Homo* and not to *P. boisei*. Considering that both of the other *Paranthropus* species are also found in deposits that yield stone tools, and at Swartkrans tools are found in deposits that only yield *Paranthropus*, the likelihood that this population had a million year tool-using heritage cannot be overlooked. Given the present evidence, however, it is not certain whether *P. boisei* used those Oldowan tools found in Olduvai alongside its fossils. There are claims that the Chesowanja deposits yielding *P. boisei* show evidence of fire use. Counting Swartkrans, this would make two *Paranthropus* deposits with such evidence.

Animals and Habitats

In addition to those animals found with *Homo habilis* at Olduvai, *Paranthropus boisei* may have additionally shared its habitat with upper Bed II animals. This includes species representing modern genera, i.e., a giant gelada baboon, a clawless otter, a white rhinoceros, a black rhinoceros, a bush pig, a warthog, a large giraffe, a large kudu, a hartebeest, a topi, a wildebeest, a sable antelope, a springbuck, and a Grant's gazelle. Two species of an extinct buffalo (*Pelorovis*) and four species of extinct pigs (*Kolpochoerus* and *Stylochoerus*) are members of extinct genera not found below upper Bed II. Finally, *Paranthropus boisei* also shared its habitat at Olduvai with *Homo*.

At Koobi Fora and Omo there are a number of animals found with *Paranthropus boisei* not present at Olduvai Gorge. Probably reflecting the older age of these deposits relative to Olduvai, a large majority of these are extinct genera, including a prehistoric baboon (*Parapapio*), various leaf-eating monkeys (*Rhinocolobus*, *Paracolobus*, and *Cercopithecoides*), a scimitar saber-tooth (*Homotherium*), a false saber-tooth (*Dinofelis*), three species of three-toed horses (*Hipparion*), three species of warthog (*Metridiochoerus*), the short-necked giraffe, (*Sivatherium*), a prehistoric kob (*Menelikia*), and the giant hartebeest (*Megalotragus*). Living genera at Koobi Fora not present in Olduvai included a mangabey, a striped hyena, a pygmy hippo relative, a waterbuck, a lechwe, a kob, an oryx, a forest duiker, a steenbuck, and a dik-dik. Overall, these animals represent a mosaic of habitats, which include small patches of forest (forest duiker and mangabey) with heavy underbrush (dik-dik and steenbuck), flooded grassland (hippo, waterbuck, lechwe, kob, and *Menelikia*), and grasslands and scrublands (oryx and *Megalotragus*). Of these habitats, *P. boisei* probably spent its time in flooded grasslands. The presence of waterbuck, kob, and lechwes at the Omo and Lake Turkana deposits, and their absence at Olduvai, probably reflects fresher water at the former sites. and the quality of graze along lakeshores and flooded river banks on which these animals depend.

An awakened *Paranthropus boisei* scours through dense grasslands in search of succulent plants and tubers for a morning meal.

Climate

Presently all of the areas in which *Paranthropus boisei* has been found have relatively long dry seasons, lasting over three months. The wettest months produce as much as 325 mm of rain and annual rainfall is below 1200 mm. Temperatures are always above freezing, and monthly highs never average over 30° C.

Classification

There would seem to be no doubt that *Paranthropus boisei* is a clearly identifiable species. Its characteristic and distinct appearance is undeniable. Skulls attributed to this species, however, all show features (e.g., face dimensions, crests, and muscle attachments) that in the living great apes and humans are characteristic of males, creating doubt as to whether females of this species can be identified. In fact, only one partial skull is claimed to be a *P. boisei* female (KNM-ER 732). The sex of this fossil, however, is by no means certain. Doubts are further compounded by the fact that there are precious few fossil bones aside from those of the skull that can be directly attributed to *P. boisei*, and what is attributed is very human-like. Because *P. boisei* females are probably smaller than males and may not show the exaggerated features of the male skull, it is likely that their fragmentary remains are more readily mis-identified as *Homo*.

Very incomplete remains assigned to *Paranthropus boisei* from Malema Valley in Malawi, Omo, and Koobi Fora could just as well represent another *Paranthropus* species. In the case of Malema, these remains are too incomplete, of an uncertain age, and too far away from other *P. boisei* remains to assume they are *P. boisei*. At Koobi Fora and Omo, teeth assigned to *P. boisei* (KNM-ER 5877 and L29-43) differ too much in size from each other to belong to the same species, and two species are likely.

The apparently close skeletal similarities between *Paranthropus boisei* and *Homo* are significant. They indicate the two shared a recent common ancestor. When more remains are found their separation into different genera may prove difficult to justify.

Historical Notes

It is not clear whether the milk canine and milk molar (OH3) found in 1955 by Louis Leakey are *Paranthropus boisei*. If they are not *P. boisei*, then Mary Leakey is credited with finding the first undoubted *P. boisei* remains, the OH 5 skull in July 1959 in Bed I Olduvai Gorge. In a *Nature* issue that same year, Louis Leakey announced the OH 5 skull discovery, coining a new binomial *Zinjanthropus boisei* for what he believed to be a human ancestor. In 1967, Phillip Tobias published his description of the OH 5 skull in the second volume of the Olduvai Gorge series (Oxford University Press). He left no doubt that this skull was closely allied to, but different from the robust australopithecines from South Africa.

In a January 1969 issue of *Nature*, Michael Day reported on what he believed to be the upper thigh bone of *Paranthropus boisei* and noted its similarity to those thigh bones from Swartkrans. In that same year, the KBS Member of the Koobi Fora Formation yielded its first *P. boisei* skull (KNM-ER 406). This skull was described by Richard Leakey in an April 1970 issue of *Nature* and was erroneously claimed to be 2.9 million years old (currently dated to 1.7 million years old). In a 1997 issue of *Nature*, G. Suwa and nine colleagues reported on the Konso *P. boisei* skull, the first found in Ethiopia.

The following institutes house the original *Paranthropus boisei* remains: National Museum of Kenya, Nairobi, Kenya; National Museum of Ethiopia, Addis Ababa, Ethiopia; National Museum of Tanzania, Dar es Salaam, Tanzania; and Department of Antiquities Lililongwe, Malawi. The species name "*boisei*" commemorates Charles Boise, the person who covered the Leakeys' expenses excavating Olduvai Gorge.

WHEN OPPORTUNITY KNOCKS

A shrill, squeaking cry caught the attention of two young men walking in the high veldt. They stopped, turned and looked around, but they saw nothing. Glancing at each other with quizzical looks they continued on their way. Then they heard the cries again. Again they stopped and turned around. This time, however, the cries lasted longer and were more forceful, enabling one of them to detect their source. He pointed to a large patch of grass out of which the edge of a dolomite boulder poked. Both walked softly with short steps as they approached the area, hoping to hear the sound again. Instead, emanating from the patch of grass they heard the sound of hooves grating on a gravely surface. When they pulled the grass back, they realized the sound was coming from a deep hole in the ground. Looking in, all they could see was the movement of horns. It appeared to be a trapped antelope.

Excitedly they ran towards the stream to look for sticks among the trees growing there. A large sagewood bush caught the attention of one of them. With the edge of dolomite block he held in his hand, he quickly stripped the branches off the trunk to make a clean wooden shaft. He fashioned one end of the shaft into a sharp point, peeling wooden fibers off one at a time. Returning to the cave, he stuck the shaft through the hole, hoping to spear the animal. The spear, however, was too short to reach the animal. Searching the area, they found another opening into the hole. The opening was too small to fit through, but prying out a boulder, they made the hole large enough so they could crawl in. After their eyes adjusted to the light they saw a large roan antelope with a broken leg lying on the ground. They speared it just above the breast plate and waited for it to die before dragging it out into the light.

Outside, they quickly butchered it, using chert they had found in the hole to make sharp flakes for cutting the meat. They ate the spleen and some of the liver, but most of the meat and bones they packed for an easy trip back to camp. There was too much meat, however, for both to carry. They would need to get help. Vultures were already flying above them. Off in the distance towards the morning sun they could see animals gathering at the waterhole. A thin column of smoke emanated from their campfire at the cave entrance. One of the men ran back to camp, while the other, spear in hand, stayed with the kill.

At the same time, more than three thousand kilometers to the north on the other side of the equator, two women foraged for rootstocks in a reed bed along a muddy lakeshore. It was six months into the dry season, the lake had retreated and there was very little food. The water line had dropped below the reed bed floor, leaving puddles of water in its wake. One of the women noticed the puddles were full of small fish. Cupping water from the puddle in both hands, she trapped two of them. After watching them swim in her hands, she brought them up to her mouth, slurping up and swallowing the water with the fish in it. Walking back to the trail, the other woman returned with a palm frond. Combing the puddle with the frond, she trapped dozens of fish in the frond fibers. Bringing the frond up to her mouth, she picked the fish off with her lips and tongue and swallowed them. By noon the two women had cleaned out all the fish in the puddles. What they hadn't eaten they piled up on large leaves. They would take these back to camp to share with the rest of their group.

Human populations now stretched the length of Africa. Living in different areas with different vegetation, animals, and climates, these humans all faced different obstacles. The opportunistic nature of humans and the ability to change their behavior to suit any given situation constantly led to new solutions for survival in dire circumstances.

HOMO ERGASTER

Skull, Teeth, and Diet

The *Homo ergaster* skulls and lower jaw bones show an amount of variation that is too great to fit in a single species. The skull (KNM-ER 1805) originally designated by Colin Groves and Vratislav Mazak as *H. ergaster* and the skull (1813), which Groves later argued is also *H. ergaster*, are both very similar to *Homo habilis*. Principally, they both show a set of skull base features associated with a face that is tucked in under the braincase and a foramen magnum that is displaced forward. The eyes are set wide apart with a single, continuous, moderate-sized bar of bone that spans above them. The top of the face is set nearly vertical, but the upper jaw juts forward considerably. Nonetheless, the roots of the upper front teeth are vertically set, and the plate of bone holding their roots is broad from top to bottom, widely separating the mouth from the nose opening.

As noted by Groves and Mazak, the braincase arches upwards from the forehead but then flattens out above. There is relatively little narrowing of the braincase behind the eye sockets. At 582 cc, the KNM-ER 1805 braincase volume is just 8 cc below the known *Homo habilis* range. The KNM-ER1813 skull has a 508 cc braincase volume, 82 cc lower than that of the minimum *H. habilis* value. The teeth are all in the size range of *H. habilis*, with the palate of the KNM-ER 1805 and KNM-ER 1813 skulls resembling that of the OH 13 and OH 65 skulls.

By contrast, the skulls of KNM-ER 3733, KNM-ER 3883, and KNM-WT 15000, later assigned to *Homo ergaster*, are much larger, with braincase volumes of 848 cc, 804 cc, and 909 cc. In the adult, the brow ridge above the eyes is more strongly formed and juts out forward. The nose bones also jut out, forming a noticeable elevation below the nose bridge. The base of the nose opening is very broad. The cheek bones are slender (narrow from top to bottom) and do not flare out towards the sides as much as do those of KNM-ER 1805 or KNM-ER 1813. The teeth are about the same size as in the KNM-ER 1805 and KNM-ER 1813 skulls, but smaller relative to skull size. The cheek tooth rows are widely spaced apart, and the upper front tooth row is relatively wide. Based on modern human tooth eruption sequences, the KNM-WT 15000 skeleton (the Nariokotome Boy) is that of a nine year old.

The narrower front tooth row of the KNM-ER 992 lower jaw bone, and the jaw's narrow width to length ratio, would appear to indicate that it corresponds to the small brained KNM-ER 1805 and KNM-ER 1813 skulls, but this is by no means certain.

There are no detailed studies of the diet of *Homo ergaster*. Similarities in teeth and jaws and muscle markings suggest a diet that was similar to that of *Homo habilis*. In keeping with the higher metabolic needs of a larger brain, the KNM-ER 3733, KNM-ER 3883, and KNM-WT 15000 individuals may have eaten a relatively greater amount of meat than the smaller KNM-WT 1805 and KNM-ER 1813 individuals. Otherwise, an eclectic diet of tubers, fruits, seeds, nuts, insects, and small vertebrates would have been the "rigueur du jour."

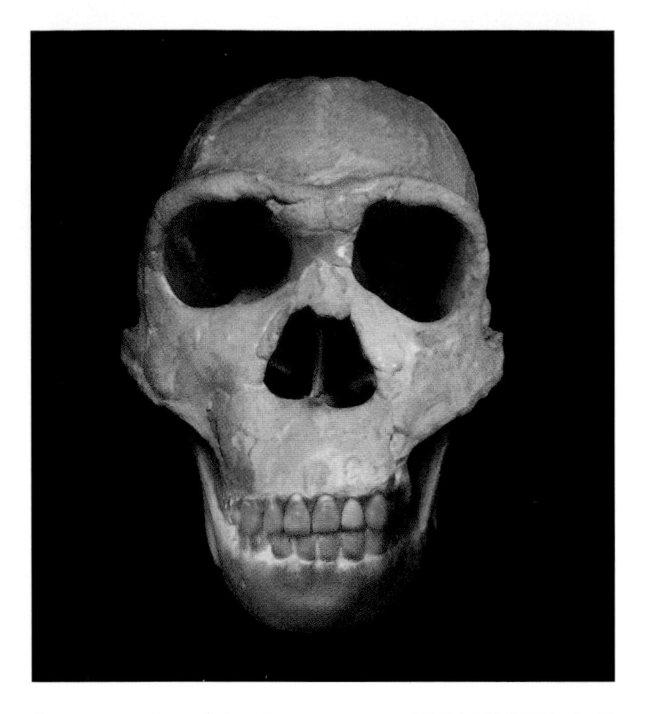

Reconstruction of the *Homo ergaster* KNM-ER 3733 skull, using the KNM-ER 370 mandible.

Skeleton, Gait, and Posture

There are no associated skeletal bones for the smaller KNM-ER 1805 and KNM-ER 1813 skulls. KNM-WT 15000 is the most complete skeleton of a possible *Homo ergaster*, preserving a nearly complete skull with a lower jaw, the pelvis, shoulder blades, backbone, and all of the limb long bones. Unfortunately, its bone growth plates had not yet fused and its joint surfaces were not fully developed. The range of movement for its joints is thus unknown. Based on growth plate closure in modern humans, the individual's age at death was twelve years old. The difference between this age estimate and that based on modern human dental eruption, which estimates death at nine years old, attests to developmental differences relative to humans.

KNM-WT 15000's stature when it died would have been 159 cm. Assuming a modern human growth curve, this adolescent could have been as tall as 185 cm when fully adult. Using modern human relationships between body weight and long bone girth and length at different ages, its weight at time of death is estimated at 52 kg, and would have reached approximately 68 kg when adult.

In all respects, the skeleton is surprisingly human-like. Limb length proportions are very much like those of modern humans except that the shin bone appears to be relatively longer than average for humans. Lower limb long-bone joint orientations also appear to be human-like and suggest thigh and lower leg movements were in the plane of forward motion. Even without range of joint motion evidence, its strong similarities to humans indicate that this species would have moved on two legs as do modern humans. It has been suggested that the proportionately longer shin bone would have been associated with more bend in the knee when walking. Differences in the shoulder blades and lower back when compared to modern humans also suggest a different use of the upper limbs. A much longer learning period of four-legged crawling than that experienced by modern human infants is a strong possibility.

Fossil Sites and Possible Range

According to its original describers *Homo ergaster* is found in the Okote, KBS, and possibly in the upper Burgi Members of the Koobi Fora Formation (the KNM-ER 1813 skull). It is also claimed to come from the lowest part of the Natoo Member of the Nachukui Formation, west Lake Turkana. Volcanic ash inter-layering the Koobi Fora and Nachukui Formations enables radioisotope dating of the deposits. Similarities between the SK 847 skull from Swartkrans Member 1 and the KNM-ER 3733 and KNM-ER 3883 skulls suggest Swartkrans Members 1 and 2 *Homo* fossils may belong to *H. ergaster*. It is likely that some very fragmentary fossils from Members H, K, and L in the Shungura Formation also represent *H. ergaster*.

Age

Found in the upper part of the Okote Member between the black pumice and Chari tufts (ash layers), the KNM-ER 992 lower jaw is between 1.49 mil-

Homo ergaster

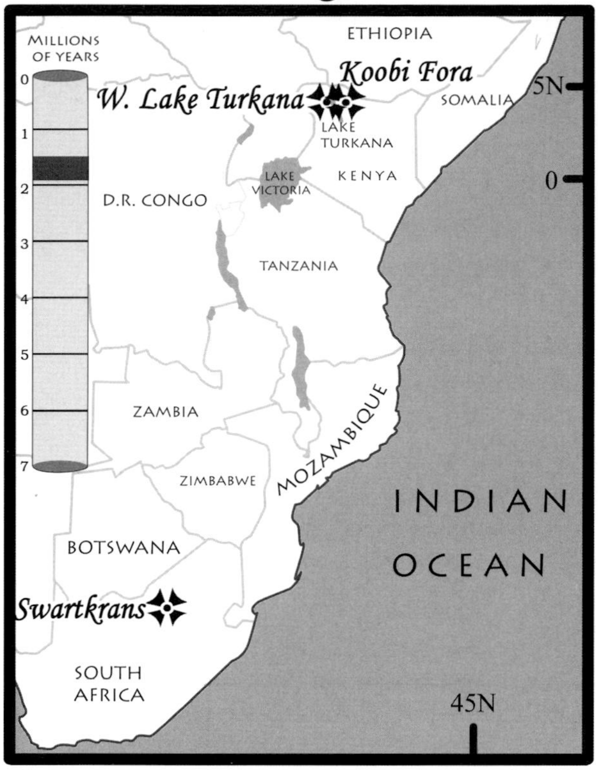

lion and 1.55 million years old and represents the youngest of the *Homo ergaster* fossils. Found in the upper Burgi Member between the KBS and Lorenyang tuffs, the KNM-ER 1813 skull is the oldest of the *H. ergaster* fossils, with an age between 1.88 million and 1.9 million years old. The KNM-WT 15000 adolescent male skeleton, the lone *Homo* fossil reported from the Nariokotome Formation, is found below the Nariokotome and Chari tuffs dated at 1.33 and 1.39 million years old, and above the Morutot Tuff dated at 1.64 million years. Assuming a constant sedimentation rate, sediment thickness deposited per year was calculated from total sediment thickness between the dated layers of volcanic ash (i.e., Chari and Nariokotome tuffs). This rate was used to extrapolate an age of 1.55 million years for the KNM-WT 15000 skeleton, based on the depth in which the fossil was found below the Chari Tuff. All *H. ergaster* fossils thus appear to span a time interval between 1.49 and 1.9 million years ago.

Tools

Both the KBS and Okote Members have yielded Oldowan-like stone tools. A much higher than expected proportion of flakes struck from the inside of stone cores, as opposed to those struck from the stone's outer surface at a number of localities, has been interpreted to mean cores were first prepared elsewhere before being carried to a final preparation site. If this is the case, the toolmakers were both transporting cores and choosing the specific material to transport. Because *Paranthropus boisei* is also found in the same deposits, it is not certain that *H. ergaster* was the toolmaker. Notably, in the upper Burgi Member where stone tools have not yet been found, *Homo* outnumbers *Paranthropus* six to one.

Animals and Habitats

Assuming *Homo ergaster* occurs throughout the upper Burgi, KBS, and Okote Members, it would have shared its habitat with a diverse number of animals, many of which today are extinct. Representing extinct genera, there was a prehistoric baboon (*Parapapio*), various leaf-eating monkeys (*Rhinocolobus, Paracolobus,* and *Cercopithecoides*), a scimitar saber-tooth (*Homotherium*), a false saber-tooth (*Dinofelis*), a prehistoric elephant with lower jaw tusks (*Deinotherium*), three extinct species of three-toed horses (*Hipparion*), three species of extinct warthog (*Metridiochoerus*), an extinct bush pig (*Notochoerus*), two prehistoric cattle species (*Pelorovis*), the short-neck giraffe (*Sivatherium*), a prehistoric kob (*Menelikia*), a giant hartebeest (*Megalotragus*), and two species of an extinct hartebeest (*Parmularius*).

Modern genera include a gelada baboon, a mangabey, a black-backed jackal, a striped hyena, a laughing hyena, a gerbil, a springhare, five rat and mouse species, an African elephant, two Indian elephant species, both the white and black rhinoceroses, two horse species, two zebra species, a hippopotamus, two relatives of pygmy hippopotamus, four giraffe species, a waterbuck, a lechwe, three kob species, two kudu species, two bushbuck species, a roan, a wildebeest, three topi species, a Hunter's antelope, an oryx, two impala species, a springbok, a forest duiker, a steenbuck, a dik-dik, and four gazelle species.

Overall, these animals were found in a mosaic of habitats, which include small forest patches (forest duiker and mangabey), heavy underbrush (dik-dik and steenbuck), and flooded grasslands (kobs, waterbuck, and lechwe). Open habitats with woodland and heavy underbrush close to water must have been very common and probably the areas used by *Homo ergaster* and the other early hominid and human species with which it shared its habitat, e.g., *Paranthropus boisei, Homo rudolfensis*, and possibly *Homo habilis.*

Climate

In both west and east Lake Turkana the climate was probably similar to what it is in Lodwar today (*see Australopithecus anamensis*). As it does today, Lake Turkana and the rivers draining into it would have supported a varied vegetation regardless of the rainfall amounts in the lake or in the river delta.

Classification

Currently there is no agreement among paleoanthropologists as to what actually is *Homo ergaster.* This is not surprising considering that the holotype specimen KNM-ER 992 lower jaw bone has also been assigned to *Homo habilis, Australopithecus,* and *Homo erectus,* and many of the other fossils claimed to belong to *H. ergaster* are too incomplete to classify to species or even genus.

If the KNM-ER 992 holotype lower jaw does not contain features that can clearly distinguish it from other *Homo* species, then *Homo ergaster* is not a valid species. The problem is that many of this jaw's features (i.e., premolar dimensions and root numbers, and cheek-tooth size), which out of necessity must be used to recognize *H. ergaster,* are too variable in humans and living great apes not to expect *Homo habilis, Homo erectus,* and possibly *Australopithecus* to also exhibit them as a matter of variation.

Assuming for simplicity's sake the most likely scenario, that the KNM-ER 992 lower jaw bone represents a progressive *Homo* species not found in the upper Burgi, then there are at least three if not four different *Homo* species at Koobi Fora: *Homo rudolfensis,* a species found only in the earlier upper Burgi Member; *Homo habilis,* found in the upper Burgi and lowest parts of the KBS Member; and *Homo ergaster,* found in the Okote Member and in the intermediate and upper parts of the KBS Member.

Some distinctive differences in the *Homo ergaster* jaw bones, however, appear to correspond with skull size. The larger skulls with greater brain volume and wide palates (e.g., KNM-ER 3733 and KNM-ER 3883 with braincase volumes of 848 cc and 804 cc) correspond to lower jaw bones (e.g., KNM-ER 730) that are broader, sturdier, and with a wider front tooth row than the KNM-ER 992 jaw. This suggests the possibility of two contemporaneous *Homo* species within the upper KBS and Okote Members—a large brained, nameless species, represented by the KNM-ER 3733 and KNM-ER 3883 skulls, and *H. ergaster,* a smaller brained species, represented by the KNM-ER 992 jaw and a recently found "small skull" from the Okote Member (KNM-ER 42700). With a 909 cc brain volume, the KNM-WT 15000 skull corresponds to the larger-brained species. Far from adulthood, it lacks the more developed brow ridge of the KNM-ER 3733 and KNM-ER 3883 skulls, and other adult facial features.

When postulating that the KNM-ER 992 jaw belongs to a small-brained individual (i.e., *Homo ergaster*), the problem of distinguishing it from *Homo habilis* persists. In this case, the KNM-ER 992 lower jaw appears as the later continuation of

One and a half million years ago a cool, gentle breeze off of Lake Turkana would have been a very welcome sensation to a male *Homo ergaster* in need of relief from the heat of the sun.

the earlier occurring *H. habilis* (i.e., the KNM-ER 1805 and KNM-ER 1813 skulls). In describing features of the KNM-ER 992 jaw bone, the creators of *H. ergaster* failed to point out distinctive differences that distinguish other fossils referred by them to *H. ergaster* (i.e., the KNM-ER 1805 and KNM-ER 1813 skulls) from *H. habilis*.

It remains to be seen whether the "small skull" from the Okote deposits will correspond to the KNM-ER 992 jaw bone and also be distinguishable from *Homo habilis*. Considering the significance of hand and foot bones in the *H. habilis* definition, associated hand and foot fossils would most surely show whether *Homo ergaster* is distinctive. For the moment, with only the KNM-ER 992 lower jaw to go on, the most practical measure would be to assign all upper KBS and Okote Member *Homo* fossils to *H. ergaster* and assume that species variation, especially between males and females, accounts for the size differences.

Historical Notes

The KNM-ER 992 lower jaw was first found in 1971. In a June issue of *Nature* the following year, Richard Leakey described the KNM-ER 992 find as *Homo*, but demurred from assigning it to a species. In a 1972 issue of *Nature*, John Robinson claimed KNM-ER 992 was indistinguishable from *Australopithecus africanus*. In 1975 in the Czechoslovakian *Journal Casopis Pro Mineralogi A Geologi*, Colin Groves and Vratislav Mazak coined the species name *Homo ergaster*, designating the KNM-ER 992 lower jaw as the holotype specimen. In 1978, in Volume I of the Koobi Fora project (Clarendon Press), Richard Leakey, Meave Leakey, and Anna Behrensmeyer referred the KNM-ER 1813 skull to *Australopithecus*. Six years later in August of 1984, Kamoya Kimeu found the Nariokotome Boy, KNM-WT 15000, skeleton. The following year in a May 1985 issue of *Nature*, Frank Brown and colleagues referred the Nariokotome Boy to *Homo erectus*. In his 1989 book *A Theory of Human and Primate Evolution* (Oxford University Press), Colin Groves included the KNM-ER 1813 skull in *Homo ergaster*. In a February 1992 issue of *Nature*, Bernard Wood placed the Nariokotome Boy in the species *H. ergaster*.

With the exception of the Swartkrans fossils, which are housed in the Transvaal Museum, Pretoria, South Africa, all *Homo ergaster* fossils are housed in the National Museum of Nairobi, Kenya. In

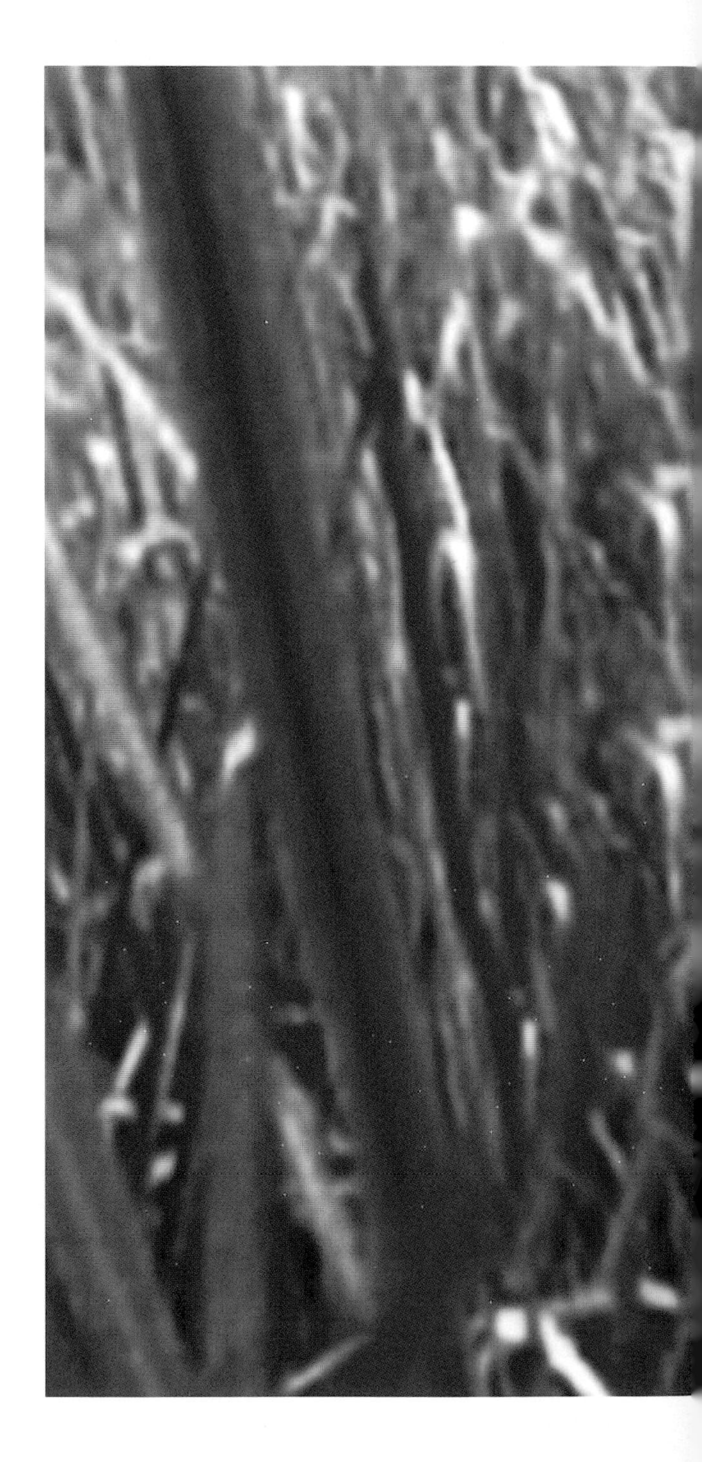

Greek, "*ergaster*" means "workman." According to Groves and Mazak, the species name was given in reference to the stone tools found in levels contemporary with the fossil. Some fossils placed in *H. ergaster*, however, come from the upper Burgi deposits where there are no stone tools.

After being separated from his group for several days, a young *Homo ergaster*, Nariokotome Boy, rejoices at seeing the familiar faces of his family.

FROM AFRICA TO ASIA?

Homo georgicus • *Homo erectus* • *Homo pekinensis* • *Homo floresiensis*

Homo fossils appear in Asia almost simultaneously with the occurrence of *Homo* in Africa. Although over the past decades there has been a general consensus among scientists that *Homo* originated in Africa, these Asian fossils and their early occurrence are challenging this wisdom.

A HANDFUL OF KNOW-HOW

She didn't remember why, but at the time she was crying. Maybe she was sick or just hungry. The old woman held her in her arms, rocked her back and forth, and hummed. She placed some cherries in her hand. Eating them made her stop crying, and she felt better. Those were the earliest memories she had. Even in these early memories, the old woman had no teeth and a scar across her cheek. Her skin was tanned, wind dried, and wrinkled. It hung loosely on her bony frame thinly covering her ribs and collar bone. Her breasts were no more than thin flaps of skin. Her arms were wiry with thick blue veins on the back of her hands and forearms. She was tiny and looked like a boy when seen at a distance from behind.

The old woman had an even disposition. Nothing made her angry or sad. She was both gentle and head strong, kind but firm. Only occasionally did she smile.

She knew of the old woman as her father's mother, someone who would help not just her or her brothers, but all the other children in the group when they were sick, in trouble, or feeling bad. It wasn't until she herself became a mother and had learned more about life that she first began to appreciate who the old woman really was, what she meant to her family, and what she had gone through in life.

She knew the old woman was once a baby that fed at her mother's breast. While still a child her group had moved to the lowlands by the sea from the desert in the south, across hilly terrain, following the animals they hunted. Because the lowlands were warm, many animals would visit during the winter, and there was always plenty of food. Every year during the rains when the weather turned warmer, the herds would disperse into the highlands to feed on the new growth. The group would stay in the lowlands until the trees bore fruit, and then follow the herds into the highlands. They would leave the highlands in the fall when the weather turned colder.

It was as a little girl that the old woman scarred her face. Climbing a fruit tree, she fell on a broken branch. Back then, there were few humans living in the lowlands. Nevertheless, she found a mate and gave birth five times. Her firstborn and her last were sons. Those in between were all daughters.

Shortly after their last child was born, her mate was trampled and killed by a wooly mammoth. It was a summer in the highlands. She never found another mate. Her oldest son took over the group for a while. He had many mates and many children. One day during a summer in the highlands, the oldest son left on a hunt and never returned. They found his body the next day partly eaten by hyenas. The old woman, and her youngest son, who in time would take over the group, managed to keep everyone together when they returned to the lowlands that autumn. Within a year the youngest son found a mate, and in the first three years his mate gave birth to two children. Some years later during winter when his mate was expecting the third child, the youngest son died.

Her memories of the old woman the year her father died were the most vivid memories that she had of her. The group was immobilized by grief and everyone appeared directionless. She, her mother, and little brother all stared into space crying. The old woman put her arms around them and shared some food she had collected. Although she must have grieved for her son, the old woman never shed a tear. When the weather turned warmer, she led the group, an assortment of women and children, to the highlands. She brought them back when the weather turned colder. That year she was the glue that kept the group together.

Now the old woman was dead. She had fallen ill during the winter and died just after making the trip to the highlands. None of her peers were there to mourn for her. She had outlived them all including her sons and daughters.

Over the years the old woman's skill, knowledge, and experience had kept the group going. She knew when to move, the trails to follow, and where to find good drinking water when there was none. She knew all about the animals and plants. She knew which plants and plant parts were edible, the preparation some plants needed to make them edible, and where the animals and plants could be found. She also knew many uses for plants, ranging from which plants healed to which could be used for construction and cord and packaging material. Fortunately for the group, the knowledge the old woman had of the land, animals, and plants, didn't die with her. It was passed-on to her grandchildren. The old woman also passed-on her habits, those everyday things she did that made everyone's life in the group easier.

She had picked up the old woman's habits, although she was not always aware of it. Whenever, her actions reminded her of the old woman, she would think of the handful of cherries and smile.

HOMO GEORGICUS

Skull, Teeth, and Diet

Homo georgicus is known from a large jaw bone (D2600), three skulls with jaw bones (D2700/D2735, D3444/3900, and D2282/D211), and one without (D2280). Representing a total of five individuals, these fossils probably belong to a large male, a sub-adult male, a toothless old female, a young adult female, and an adult male. The skulls are relatively small, with an average 650 cc brain size, and a 610 cc to 775 cc range. The skull vault is elongated, narrow, and low (short from top to bottom), and shows a decrease in width with increasing distance above the ears. When the skull is viewed from behind, it presents a pentagon-shaped outline as seen in *Homo erectus*. At the top along its midline, the braincase is thickened, forming a keel. Behind the eye sockets, the braincase is constricted. The single brow ridge that spans across and above the eye sockets is not as heavily thickened as in *H. erectus*. When compared to *H. erectus*, the *H. georgicus* face is smaller, but the upper jaw and teeth are more forward protruding.

The teeth are human-like in most respects, smaller than *Australopithecus*, and close in size to the mean of African *Homo* fossils of comparable geologic age. The D2600 jaw bone is an exception in that it is larger than that of known *Homo erectus*, *Homo habilis*, *Homo ergaster*, and *Homo rudolfensis*. The canines are also larger in the D2600 jaw bone and the premolars have double roots as opposed to the single rooted premolars common to D211 and most *Homo* species. The first premolar in the smaller D211 jaw bone shows detailed similarities to those of *H. rudolfensis* (KNM-ER 1802) and the early *Homo* from Malawi (UR 501). More than likely, differences between D2600 and the other Dmanisi remains are in part due to sex differences and to at least a forty thousand-year time interval between the two. Notably, the D2600 jaw bone corresponds in size with the D2280 skull. Considering no other primates have been found in the Dmanisi ash deposits, it is unlikely that D2600 and the other Dmanisi remains represent different species. Overall, the *Homo georgicus* skull shows a mix of early and late *Homo* features, with a face reminiscent of *H. habilis*.

Very little is known of the diet of *Homo georgicus*. The notion that movement out of Africa was fueled by *Homo* hunting and scavenging adaptations, however, has sparked much speculation. The *H. georgicus* diet, in fact, was probably not much different from that of *Homo habilis* and *Australopithecus africanus*. As a stone wielding tool user, it probably consumed hunted or scavenged meat especially in winter, but its main staples were probably tubers, roots, seeds, herbs, and, when available, fruit. Unfortunately, no primates are living in the area from which to draw direct analogies. Tooth enamel studies show that the D211 mandible individual during childhood suffered from malnutrition, suggesting that food supplies were not constant and markedly seasonal.

Skeleton, Gait, and Posture

Ribs, backbones, a collar bone, a shoulder blade, two arms bones, a thigh bone, a knee cap, a shin

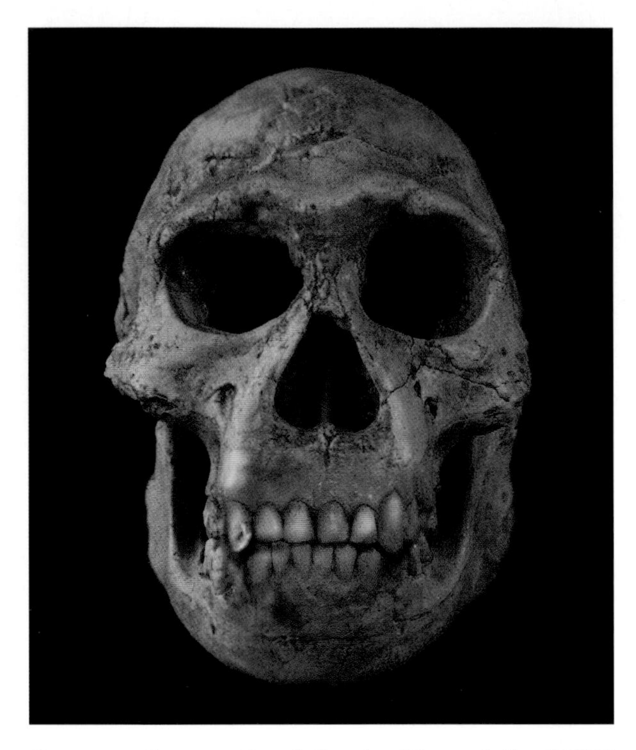

Reconstruction, mostly of the dentition, of the D2700 *Homo georgicus* skull from Dmanisi, Republic of Georgia.

bone, an ankle bone, and some foot bones of *Homo georgicus* have been unearthed and photographed. Unfortunately, these remains have yet to be formally described. So far all that has been reported on is a partial foot bone (third metatarsal). This bone is smaller but stockier than that of modern humans. It does not appear to be as compressed from side to side as the modern human bone is. Unlike the modern human bone, it has a large tubercle at its base. Although these features could indicate a less developed foot arch than in modern humans, there is too little evidence to be certain. Based on the foot bone and assuming human body segment proportions, the height of *H. georgicus* was calculated to be 150 cm. This value would be an overestimate, if it turns out *H. georgicus* has more *Australopithecus*-like body proportions, and legs shorter than modern humans. The latter seems highly unlikely given the human nature of its limb bones and foot bones as seen in published photographs.

Fossil Sites and Possible Range

Homo georgicus is found in sand and volcanic ash deposits under the medieval (ninth to fourteenth centuries) town of Dmanisi in southeastern Georgia. Dmanisi is located in the lesser Caucasus on a basalt (dolerite) hilltop 80 m above the Masavera and Pinezaouri river union, about eighty kilometers southwest of Tbilisi, Georgia. Slow moving water deposited the sand and ash layers up to a depth of 4 m on top of a basalt flow. A thin limestone crust (30 mm to 40 cm thick), resulting from underground water action after deposition, divides the deposits into lower and upper layers A and B. The top of layer A just below the limestone crust reflects a time interval in which there was some erosion and no deposition. This period of erosion formed gullies within the A deposits that were filled in by the overlying B layer deposition. Sand and ash contemporaneous with the B layer, therefore, may rest directly on the basalt flow at the same depths as the A layer. Most of the human remains were found in a gully filled during B layer deposition. The foot bone and the large jaw bone (D2600) were found in the A layer. Because fossils were deposited by relatively slow moving water and moved very short distances, they are well preserved, but they are fragile and not heavily mineralized.

Age

The volcanic origin of the deposits provides precise dating. The underlying basalt gives a radioisotope date of 1.85 million years. The overlying A layer yields a date of 1.81 million years. The B layer above the limestone crust has a reverse magnetic polarity corresponding with a reversal in Earth's magnetic pole dated to 1.77 million years ago. Found in ash with reversed polarity, all of the skulls and the 1991 jaw bone (D211) are thus slightly younger than 1.77 million years. The large jaw bone (D2600) and foot bone, found in the A layer, are 1.81 million years old. The possibility that the large jaw bone came from a more recent infilling of a gully in the A layer makes this date tenuous.

Homo georgicus

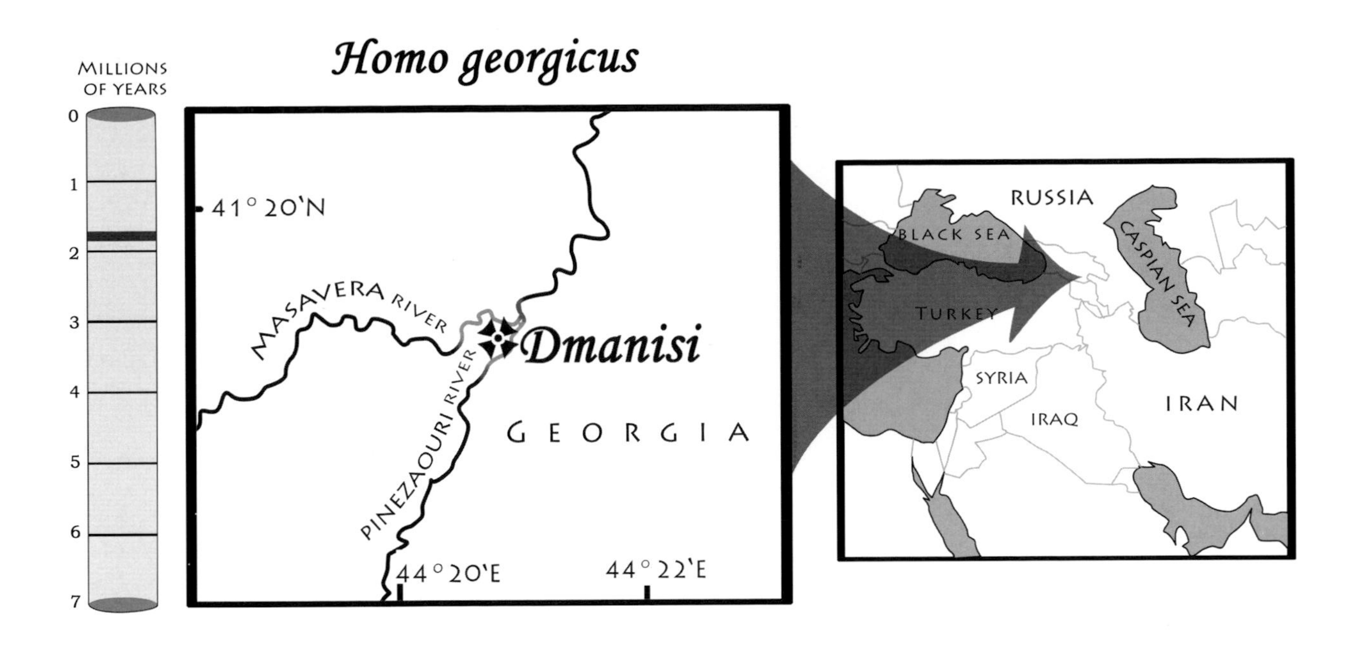

Tools

Stone tools are found in both the A and B ash layers. They consist mostly of flakes, some scrapers, and choppers. All tools are made of the local basalt and reflect a simple tool technology of flakes worked on a single side. No fossil bone so far recovered is reported to have tool marks or breaks characteristic of processing by humans.

Animals and Habitats

Fossil animals found in the *Homo georgicus* deposits are typical of temperate climates. Instead of the variety of antelopes common in the African hominid deposits, there are cattle, musk ox, and several deer species. Instead of spring hares and other typical African rodents, there are woodchucks, hamsters, gerbils, voles, and field mice. The deposits also yield gazelles, a prehistoric giraffe, horses, wooly mammoths, steppe rhinoceros (Dicerorrhinus), martens, wolves, bears, hyenas, lions, two extinct species of saber-tooth cats (the dirk-tooth and the scimitar), and an ostrich. Significantly, there are no monkeys. The ostrich, hyena, and prehistoric giraffe, all animals with an African origin, point to possible dispersal routes that may have also been followed by *H. georgicus* or its ancestors.

Overall, fossil animals reflect a number of habitats dominated by open grasslands. Pollen grains found in the deposit suggest mountain forests of firs, pines, and birch trees. Elms, hackberries, willows, alders, and linden (basswood) trees contributed to forming deciduous woodlands with thick underbrush along streams, floodplains, and wherever water was available. These deciduous trees do not survive temperatures below -29° C (-20° F), providing a minimum temperature for the climate *Homo georgicus* endured.

Climate

At forty-one degrees North latitude and 1000 m above sea level Dmanisi today enjoys cold dry winters and relatively low yearly precipitation. Tbilisi, a city 80 km northeast of Dmanisi and about five hundred meters lower in elevation, reports an annual average precipitation of 500 mm, with snow falling during the winter months. More than half of the yearly precipitation occurs in April-July and less than 56 mm fall in December-February. January is the coldest and July the warmest month, with average temperatures of 1.7° C (35° F) and 25° C (77° F). During December-February average low temperatures are always below freezing and no days have temperatures above 18.3° C (65° F).

Freezing temperatures may occur from October-April. At a higher elevation Dmanisi would have average temperatures 3.25° C lower than Tbilisi. The trees and all of the animal fossils from the Dmanisi deposits, especially the rodents, suggest the climate was not much different when *Homo georgicus* inhabited the area than it is now. The climate is similar to that of the South African cave sites (*see Australopithecus africanus*, Climate) except that at Dmanisi it rains less and a bit earlier in the spring. Two hundred kilometers west of Dmanisi on the Black Sea coast, the temperatures are somewhat warmer, so the possibility that humans visited Dmanisi only during warm weather should not be discounted.

Classification

With the exception of the D2600 jaw bone that persists as the holotype specimen of *Homo georgicus*, all other fossils have recently been referred to *Homo erectus*. Confusingly, this has been done by two of the same authors who four years earlier, based on the same fossils, erected the new species *H. georgicus*, without explaining why they changed their opinion. Considering that the holotype of *H. erectus* may be one million years younger and from a Southeast Asian island (Java) more than 9,000 km away, it is extremely unlikely that the Dmanisi fossils are *H. erectus*. It is just as unlikely that the large 2600 jaw bone from the same locality and supposedly 40,000 years older than the other remains represents a separate species (*H. georgicus*). Until more is known of the skeleton, it is much more prudent to refer all Dmanisi human fossils to *H. georgicus*. Further finds may show if the Dmanisi humans belong to two species.

More than likely additional finds will probably show that differences between *Homo georgicus* and the contemporaneous African *Homo ergaster* are restricted to minor skull differences and do not merit species distinction. If this proves to be the case, all the Dmanisi remains would be referred to *H. ergaster*. Erk-el-Ahmar, a 2 million-year-old archaeological site in Israel, two-thirds of the way from the east African rift to Dmanisi, suggests that a stone wielding *Homo* may have had a continuous range from the African rift to the Caucases.

Historical Notes

The first remains of *Homo georgicus*, an adult jaw bone with all its teeth (D211), were found by Antje Justus in September of 1991 and described by Leo Gabunia in a 1992 issue of the *Jahrbuch der Römisch-Germanisches Zentral-Museum Mainz*. Based on this original jaw bone (D211), and two partial skulls (D2280 and D2282), one foot bone, and a larger jaw bone found in September 2000 (D 2600), Leo Gabunia, Marie-Antoinette de Lumley, Abesalom Vekua, David Lordkipanidze, and Henry de Lumley described a new species, *H. georgicus*, in a 2002 issue of *Comptes Rendus de Paleovol*. The large D2600 jaw bone was named as the holotype specimen, and the species was claimed to show marked size differences between males and females. In a 2006 issue of the *Journal of Human Evolution*, Philip Rightmire, Vekua, and Lordkipanidze placed all the fossils, with the exception of the 2600 jaw bone, into *Homo erectus*, supporting an earlier classification based on the D211 mandible by Gabunia and others. The Dmanisi fossils are housed in the Department of Geology and Paleontology at the State Museum of Georgia in Tbilisi, Georgia. The species was named after the newly independent country of Georgia.

Pausing by the river's edge far from his homeland, *Homo georgicus* finds himself in a new world.

EVE WITHOUT ADAM

The clouds had been building since noon. By the afternoon, they could no longer hold the moisture, and the sky gave way to a torrential downpour. The rain lasted less than an hour and cleared the humidity. Everyday it came at about the same time. Everyday the humans would find shelter at the entrance of one of the many shallow caves on the slopes of the volcano. When the rain ended, they emerged to feed in the lush vegetation that grew on the slopes.

Upslope from the cave, a grove of rose apple trees were in fruit, attracting the majority of the group. One of the men bypassed the grove, walking downslope instead. The smell of the clear air and wet soil and the sight of the rose apple grove reminded him of another time he had walked here. The young man remembered a durian tree a little farther down where the ground leveled and water accumulated. He had left a hammerstone at the base of the tree to open its fruit, but from where he was standing it appeared to no longer be there. As he reached the tree, he saw that the hammerstone was now wedged between the tree trunk and a lateral branch. All the fruit that could be reached while standing below the tree had already been taken. Thorny husks littered the ground below. The man had no choice but to climb the tree. He picked and dropped down a dozen fruits before he descended. As he did so, he noticed a young woman below the tree, gathering the fruit he had dropped into a pile. He stared at her, noticing that her skin was smooth and clear and her breasts were full. He starred at her. When she finally looked back at him, he arched his eyebrows and smiled, showing his teeth. She reciprocated by doing the same.

Both the man and woman sat down below the tree. He rapped on the husk of the durian fruit with the hammerstone, cracking it open to get at the pulp inside. Both shared its contents, alternately bringing the open husk laden with pulp up to each other's mouths.

A group of four humans arrived at the durian tree, but the couple had their backs turned to them. One of the men in the group cleared his throat, emitting a low deep "hmm." Somewhat startled, the young man and woman turned around to look. They acknowledged the presence of the four by emitting their own deep throated "hmm." Shortly thereafter, they walked off together upslope through the bamboo forest, leaving the hammerstone for the others to use.

In the middle of the night, it rained again. The couple sought shelter by crouching below some thick shrubs. In time, the shrubs became saturated with rain, and the couple became cold and wet. Face to face, they hugged each other to keep warm. Despite their physical discomfort, both the man and woman were content. She had left her group several days ago and having a man next to her made her feel protected and cared for. He felt secure and at peace at the prospect of having a healthy woman with the physical promise of child bearing and rearing success.

After an early morning meal of bamboo spears, the couple continued their walk upslope. Just above the forest, they happened upon a clearing with few trees, but densely covered in bushes and vines. Whortleberry, raspberry, and blackberry fruit abounded. Flattening the tall grass and herbs, the couple sat eating fruit. Movement in the foliage some distance down-slope a short while later proved to be a chalicothere digging up vegetation. The rustling of leaves and the squawking and wing flapping of a jungle fowl from behind a bush close to where they sat did not alarm them. A tiger, however, had positioned itself upslope from the couple and was now crouched and ready to pounce. The man, possibly sensing its presence, turned around and screamed at the sight. The tiger pounced, knocking the couple over and clawing at both of them. Shielding the young woman, the man gave her a chance to escape by becoming the tiger's next meal. With her mate gone she no longer felt protected or cared for and now needed to be with other humans. Ultimately, those feelings of protection and security the two had shared were only an illusion.

HOMO ERECTUS

Skull, Teeth, and Diet

Javanese fossil skulls and teeth currently assigned to *Homo erectus* show an amount of variation that cannot be sensibly placed in a single species. Nearly all remains consist of braincases, with few teeth, faces, and lower jaws preserved. Preserving a face and five teeth, the Sangiran 17 skull is the most complete of the reported finds. Its face is vertically set, with its upper jaws barely jutting out past the plane of the face. The plate of bone holding the roots of the front teeth is also vertically set and widely separates the mouth from the triangular shaped nose opening. The lower edge of the eye sockets is set low relative to the nose opening. The cheek bones are very wide from both top to bottom and side to side. They attach low on the upper jaw and far forward at the level of the first molar. Above the eye sockets, there is a thick continuous bar of bone (brow ridge), which on either side juts out sideward past the eye sockets. Because the forehead does not rise up vertically, the brow ridge is not guttered above where it meets the braincase. In all the above respects, *H. erectus* differs markedly from the Chinese *Homo pekinensis*.

As in *Homo pekinensis*, the *Homo erectus* skull case is low and long, moderately narrowing in width behind the eye sockets and protruding bony brow. As in *H. pekinensis*, it also presents a keel on top of the braincase along its midline. Braincase width from side to side decreases considerably with decreasing distance above the ear. The *H. erectus* skull base, however, is much wider and not as strongly inclined backwards as in *H. pekinensis* and is thus lower, or shorter from top to bottom. A 931 cc mean braincase volume for six Sangiran individuals is close to the 941 cc value reported for the Trinil II braincase, the 870 cc for the Ngawi braincase, and the 986 cc mean for the Sambungmacan braincases. All are less than the mean for *H. pekinensis*. Ngandong fossils, however, have a much higher braincase volume (1149 cc), and the Mojokerto Child a much lower one (575 cc) than the other Javan fossils. In the older *H. erectus* fossils from the Pucangan Formation (Sangiran 4), the bone that forms the wall of the braincase is even thicker than in *H. pekinensis*. This is not the case for Trinil and the younger Sangiran fossils.

As seen in the Sangiran 4 skull, the palate is very broad, with cheek tooth rows that converge towards the front only slightly. The front teeth are thus set nearly at right angles to the cheek tooth rows. Palate and teeth are overall much larger than they are in *Homo pekinensis*, although the palate retains a human-like length to breadth proportion. As in other *Homo* species, the *Homo erectus* molars wear flat. Relative to their breadth and length dimensions, however, the *H. erectus* cheek teeth are not as high-crowned as in *H. pekinensis*. Too few associated teeth have been reported to gauge differences between front tooth, premolar, and molar size relationships, or differences between males and females.

There is no direct evidence as to the *Homo erectus* diet. The low crowned molars relative to *Homo pekinensis* suggest the diet was composed of less abrasive foods. Larger cheek teeth suggest *H. erectus* had much more roughage in its diet. This is in accord with the lush vegetation and year-round fruit availability characteristic of tropical climates with high rainfall. With a short dry season it is unlikely *H. erectus* would have consumed considerable quantities of animal meat, although large insects and small vertebrates would have been eaten occasionally.

Skeleton, Gait, and Posture

Only a few *Homo erectus* the lower limb bones are known. From Trinil there is the complete thigh bone from the *H. erectus* holotype specimen (Trinil 3) and six thigh bone shaft fragments. One of the latter (Trinil 6) is nearly complete and missing only the joint surfaces. There are two thigh bone shaft fragments from Sangiran (Sangiran 29 and Sangiran 30) and two from northwest Sangiran (Kresna 10 and Kresna 11). From Ngandong there is a fragmentary pelvis and two shin bone shafts (A and B), and from Sambungmacan, a shin bone shaft fragment (SM 2).

Except for the complete *Homo erectus* holotype thigh bone and the Ngandong shin bone B (Ngandong 14), none of the other fossils preserve joint surfaces. The size and structure of the complete thigh bone is identical to that of modern humans, indicating that its bearer must have moved around in the same fashion as modern humans. For instance 1) the thigh bone's mid-shaft cross-section is small relative to the its large ball-like joint surface of thej hip joint, 2) the angle formed between the thigh bone's neck and shaft is approximately one hundred and twenty degrees, close to the human mean, 3) there is a nine degree carrying angle formed between the thigh bone's shaft and its knee joint surface, 4) on its backside along its length the thigh bone shaft has a longitudinal ridge of bone for muscle attachment, 5) the thigh bone's mid-shaft x-section is tear-drop shaped, and 6) its joints are oriented so the kneecap faces forward when the hip is extended. Unlike many supposed fossil *Homo* thigh bones (from Koobi Fora, Olduvai, and Zhoukoudian), it does not show the pronounced front to back flattening of the shaft (platymeria). All the other known Javanese thigh bone shafts are identical to the complete Trinil thigh bone, confirming the human-like structure of the *Homo erectus* thigh bone. The described shin bone shaft fragments are like those of modern humans, but more robust. Assuming human-like proportions and based on the two most complete thigh bones (Trinil 3 and Trinil 6), the height and weight of *H. erectus* is estimated to be 163 cm and 54 kg. The bearer of the Ngandong shin bone must have been heavier, although its height may have been comparable.

Fossil Sites and Possible Range

Homo erectus is known from lake and river sediments associated with volcanic activity at the following localities: Sangiran, Sambungmacan, Trinil, Ngawi, Ngandong, Kedung Brubus, and Mojokerto (listed from west to east), on the eastern half of the Island of Java, Indonesia. Volcanic activity has uplifted the sediments, and the Solo River and its tributaries have cut through them, exposing them at various localities. At Sangiran the *H. erectus* sediments are grouped in two formations, the underlying Pucangan, also called the Sangiran Formation, and the overlying Kabuh, also called the Bapang Formation. The Pucangan consists mainly of lake sediments. It lies on a volcanic breccia, which in turn overlies marine sediments (the Kalibeng Formation). The volcanic breccia represents a mudflow from active volcanoes and contains no fossils. At the time of its deposition, volcanic activity must have just lifted this area of the island above sea level. The Kabuh Formation consists mainly of stream and river sediments. It is separated from the Pucangan by the "Grenzbank," a layer of limestone cemented pebbles and sands. This limestone contains the shells of unicellular marine organisms. The Grenzbank is considered to form the base of the Kabuh Formation. It is usually less than a meter thick, and may contain *H. erectus* fossils. Most vertebrate fossils, including *H. erectus*, however, come

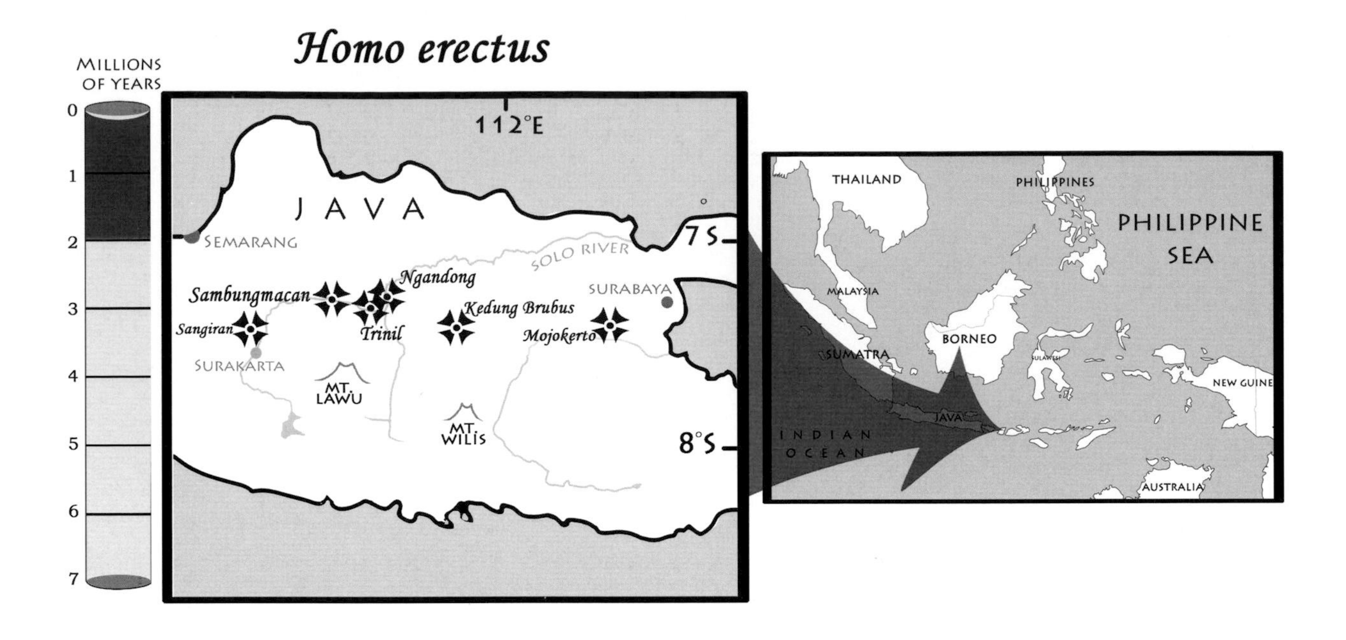

from just above or below the Grenzbank. Along rivers and streams, the Kabuh Formation is occasionally overlaid by more recent river deposits that may also contain fossils. These fossils may have a recent origin, or they may have been eroded out from the Kabuh or Pucangan Formation and re-deposited by water.

The Pucangan and Kabuh Formations do not exist at all localities. Due to faulting, outcrops at different localities have been difficult to correlate. All of this has played havoc with relative and absolute dating of fossils despite the fact that volcanic sediments allow radioisotope dating. For these reasons, the dates have large degrees of error and many are still under discussion.

Age

At Sangiran volcanic sediments analyzed from the contact of the Kabuh and Pucangan Formations yielded a date of 1.5 million years. Volcanic ash layers above all the Sangiran *Homo erectus* fossils yielded a date of 1.02 million years. All the Sangiran fossils found within the Grenzbank and in the Kabuh Formation are thus bracketed between these two dates. A date of 1.66 million years for volcanic sediments believed to come from the Pucangan Formation, 2 m above the oldest *H. erectus* fossils (Sangiran 27 and Sangiran 31), appears to correspond to the volcanic breccia underlying the Pucangan Formation and is thus questionable. This breccia has also produced a date of 1.77 million years. Given this date as the maximum for the underlying breccia, all of the Pucangan Formation at Sangiran and the fossils found in it must be between 1.5 million and 1.77 million years old. This age is corroborated by paleomagnetic studies and relative dates based on unicellular marine organisms (i.e., foraminafera).

The Trinil fossils are believed to correspond to those from the lower Kabuh, with an age between 1.0 million and 1.5 million years. Absolute dates using different techniques for middle and upper Kabuh sediments have yielded minimum ages of 700,000 years. If this dating is accurate, the Trinil fossils must be between 700,000 and 1.5 million years old. The earlier date, however, may be an overestimate if Trinil fossils also come from more recent river deposits that overlie the Kabuh as has been claimed.

Crafting stone tools by a fireside, *Homo erectus* prepares weapons for tomorrow's group hunt.

Fossils from Kedung Brubus have no absolute dates, but have generally been considered to be contemporaneous with those from the Kabuh Formation. They thus span the same possible ages as Trinil. The Ngandong *Homo erectus* skulls come from river deposits overlying the Kabuh Formation. Direct dating of bone and teeth found in the Ngandong deposits have yielded dates between 27,000 and 100,000 years. Because the Ngandong deposits have yielded a number of extinct animals, there is reason to believe this date is too young, and the Ngandong fossil skulls may be 120,000 years old or older.

At the Sambungmacan *Homo erectus* locality, neither the Pucangan nor Kabuh Formation can be identified. The fossils come from sediments directly overlying the Kalibeng (marine) Formation. Paleomagnetic studies show that Earth had a normal polarity when the *H. erectus* sediments were deposited, corresponding to an age of less than 780,000 years or one between 990,000 and 1.07 million years. These fossils thus have a maximum age of about one million years.

At Mojokerto, absolute dating of volcanic ash and rocks in deposits from where the child's skull was believed to be found yielded dates between 1.81 million and 1.9 million years. Unfortunately, it is not certain where the skull came from. A recent study places the Mojokerto skull in deposits 20 m above the dated volcanic ash layer, indicating it is younger than 1.81 million years. The Ngawi *Homo erectus* skull cap was found on the surface so it is not known if it comes from the Kabuh Formation or from an overlying river deposit. Considering all Javan sites, *H. erectus* fossils span a possible time interval between 27,000 years and 1.9 million years.

Tools

There are no tools found at the Kedung Brubus, Mojokerto, Ngawi, and Trinil *Homo erectus* localities. Crude flakes struck on a single side from coarse stones have been reported at Sangiran, although these are claimed by some not to be tools. Ngandong and Sambungmacan *H. erectus* deposits have both produced tools consisting of cores, choppers, and un-retouched flakes from coarse volcanic rocks. None of these tools have been worked on both sides.

Animals and Habitats

Classically, there have been three different animal assemblages associated with *Homo erectus* deposits from Java: Ngandong, Trinil, and Jetis. Spanning a time interval of nearly two million years, these animal assemblages represent Upper, Middle, and Lower Pleistocene age (1.8 million to 11.6 thousand years ago). Because 1) fossil collections from Java may lack accurate locality data, 2) horizontally discontinuous deposits with similar-appearing sediments have been assumed by past scientists to represent the same geologic event and thus age, and 3) the Solo River has re-deposited older fossils in younger sediments, there is very little certainty as to which animals coexisted with which *H. erectus* fossils. Confusion as to time intervals in which fossils occur is further compounded by confusion as to the range of habitats the fossils represent. There is thus very little agreement as to possible changes in habitat over time. In more recent deposits, the occurrence of orangutans, gibbons, siamangs, leaf-eating monkeys, binturongs, palm civets, sun bears, and Asiatic mice deer all point to tropical forest conditions. Bamboo rats indicate sizeable bamboo forests. The presence of a giant pangolin, 2 m in length, reflects a bountiful supply of ants and termites. These insects occur in high concentration in tropical forests with lots of leaf litter-accumulated during the dry season.

Many other animals found in the deposits may use forests for shelter, but depend on grassy fields and shrubbery to feed, e.g., tapirs, barking deer, axis deer, red deer, rusa deer, cattle, black bucks, macaques, two pig species, hares, rats, and porcupines. Indian elephants, bears, wolves, lesser oriental civets, oriental civets, leopard cats, and two species each of dholes and of tigers are also found in the *Homo erectus* deposits. Although these latter animals may show a preference for sheltering in forests or woodlands, they are not restricted to any one habitat. A few animals indicate flood plains, shallow lakeshores, and/or swamps. These animals include two species of one-horned rhinoceros, the otter, the hippopotamus, and the Asian water buffalo.

Finally, there are also animal groups that are now extinct, including a scimitar cat (*Homotherium*), a *Stegodon* (see *Homo floresiensis*), two Asiatic cousins of the African eland (*Epileptobos* and *Duboisia*), one of which (*Duboisia)* was endemic to Java, a chalicothere (*Nestoritherium*, a plant-eating extinct cousin of rhinoceroses and horses, with large claws instead of hooves.) Many of the animals represented by two species (tiger, dhole, rhinoceros, and pig) reflect changes over time. Unfortunately, at the moment there is no agreement as to which species are older and which are younger. Overall the fossil animals suggest tropical forests with grasslands in clearings, and around swamps or shallow lakes at headwaters, deltas, or bends in watercourses.

Climate

Surakarta, Indonesia (107 m above sea level), 10 km south of Sangiran, and 54 km, 70 km, 76 km, 81 km, and 112 km west-northwest of Sambungmacan, Trinil, Ngawi, Ngangdong, and Kedung Brubus, respectively, reports nearly constant year round temperatures characteristic of a tropical island climate. The coldest mean monthly temperatures (27.8° C, 82° F) are recorded for January-March and July-August, and the warmest for October-November (29.5° C, 85° F). Annual precipitation is 2182 mm, with December-March receiving the most rainfall (1339 mm), and June-September receiving the least (185 mm). Over a forty year period Surabaya (3 m above sea level), on the Javan coast 200 km to the east-northeast of Surakarta and 36 km northeast of Mojokerto, reports the same mean monthly temperatures, but with less rain (1471 mm). Rain distribution over the year, however, was similar (December-March 1016 mm, and July-October 104 mm). In both areas rainfall amounts can support tropical forest growth. There is no evidence that suggests the climate would have been much different during *Homo erectus* times. On the contrary, many of the animals found in the *H. erectus* deposits are tropical and likely forest based.

Classification

Spanning nearly two million years and with as much as two-fold differences in braincase volumes between individuals from different localities and formations, it is highly questionable whether the fossils from Java belong to one species. The *Homo erectus* holotype specimen, consisting of the partial Trinil II braincase, has too few features to confidently compare it to any other fossils. The many features assumed to be diagnostic of the *H. erectus* skull may be a consequence of brain size and not distinctive of any one species. Except for the thigh bones, other remains assigned to *H. erectus* by Dubois (the Kedung Brubus jaw bone and the three teeth from Trinil) are in themselves unclassifiable to species. In fact, without more accurate dating it may prove difficult to associate Trinil to the more complete and plentiful remains from Sangiran.

As is, at least three species of *Homo* can be recognized in the Java deposits: 1) a relatively recent species known from between 27,000 and 120,000 years plus, characterized by a relatively large braincase volume as typified by the Ngandong skull, 2) a species occurring between 1 million and 1.5 million years in the Kabuh Formation, associated with a relatively smaller brain size (930 cc) as seen in the Trinil, Sangiran 17, Ngawi, and Kedung Brubus fossils, and 3) a species occurring in the Pucangan Formation between 1.8 million and 1.5 million years with a small braincase, thick braincase walls, and very large teeth as seen in skulls from Mojokerto and Sangiran. These three species all have available names: *Homo soloensis* for the Ngandong skulls; *H. erectus* for Trinil, Kedung Brubus, and 1-million- to 1.5-million-year-old fossils from the Kabuh Formation at Sangiran; and *Homo modjokertensis* for the older remains from the Pucangang Formation and Mojokerto.

The Sambungmacan braincases appear to be progressive in their features and intermediate between Ngandong and those from the Kabuh Formation at Sangiran. Although it may be attractive to perceive the three species as part of an evolving lineage, this may not have been the case. During this two million-year period, violent volcanic activity on Java and intermittent continuity of this island with the Asian mainland due to lowering of sea levels (*see Homo floresiensis*) caused extinctions of endemic forms and brought new immigrant populations to the island from the mainland. Unfortunately, the Asian mainland record is as yet too poor to know how the Javan human species relate to the mainland ones.

Considering what is known of the skeletons of the two more recent species (*Homo soloensis* and *Homo erectus*), their fossils are best referred to the genus *Homo*. Whether or not *Homo modjokertensis* should be referred to a different genus (i.e., *Meganthropus*) must await analysis of more complete fossils.

Historical Notes

Eugene Dubois is credited with unearthing the first remains of *Homo erectus*, a braincase at Trinil in October of 1891. It is now not certain whether the Trinil I molar he had unearthed a month earlier is a hominid. In an 1892 issue of *Verslagen Mijnwezen, Batavia*, Dubois designated the molar, braincase (Trinil II), and thigh bone (Trinil III) as the holotype specimen of a new genus and species he named *Anthropopithecus erectus*. Two years later in an 1894 paper, he changed the generic name of this species to *Pithecanthropus erectus* and claimed it to be a transitional form leading to modern humans.

Contemporary opinions varied widely from attributing the thigh bone to a giant gibbon (e.g., Hans Virchow and Johannes Bumüller) to agreeing with Dubois that it was human (e.g., Leonce Manouvrier and David Hepburn). In a 1924 issue of the *Proceedings Konlijke Acadamie Wetenschappelijke*, Dubois referred the Kedung Brubus jaw to *Pithecanthropus erectus*. In a 1932 *Wetenschappelijke Mededeelingen-Dienst van den Mijnbouw in Nederlandsch-Oost Indië* issue, Willem Oppenoorth coined the name *Homo soloensis* for the first four skulls found by Gustav ("Ralph") von Koenigswald and Carel ter Haar at Ngandong the previous year. In a 1936 issue of the *Proceeding of the Academy of Sciences, Amsterdam*, von Koenigswald made the Mojokerto Child's braincase the holotype specimen of a new species, *Homo modjokertensis*. Franz Weidenreich, in a 1945 issue of the *Anthropological Papers of the American Museum of Natural History*, referred the Sangiran 4 skull and upper jaw to a new species, *Pithecanthropus robustus*, and a partial lower jaw bone to *Pithecanthropus dubius*. In this publication he also described a new species and genus, *Meganthropus palaeojavanicus*, a name which was proposed but not formalized in von Koenigswald's correspondence and museum labels on the fossil jaw bone. In 1950, von Koenigswald, in the *Proceedings of the International Congress of Geology, London*, transferred Sangiran 4 to *Homo erectus modjokertensis*. That same year Ernst Mayr's influential paper at the Cold Spring Harbor symposium swayed paleoanthropologists to place all supposed ancestral humans into *Homo*, including *Australopithecus*.

Weidenreich's posthumous and unfinished work on the Ngandong fossils, published in 1951 in the *Anthropological Papers of the American Museum of Natural History*, suggested their relationship lay with Australian Aborigines. In 1973 Michael Day and Theya Molleson, in a special human biology issue of the *Symposia for Social Sciences*, claimed it was unlikely the Trinil femur belonged to *Homo erectus*, although their chemical analysis on the Trinil thigh bone and skullcap failed to show the two came from different deposits. In a 1980 Yale University Anthropological paper, Albert Santa Luca assigned all Javan *Homo* fossils to *Homo erectus*.

The Javan fossils are housed in Seckenburg Museum, Main, Germany; National Natural History Museum, Leiden, Netherlands; and in a number of institutions on the island of Java. The species name "*erectus*" refers to its upright, two-legged stance. The genus name first given by Dubois to the Trinil fossils, "*Pithecanthropus*," is a compound Greek word meaning "ape man" (pithecus = ape, and anthropus = man).

A WINTER NIGHT'S DESPERATION

The sky was as clear as the night was cold. The twinkling stars and the bright glow of the milky way illuminated the night, their light making up for an absence of the moon. At a cave entrance a group of humans crowded around a fire. A tapestry of animal skins was hung from one side of the cave entrance with the hope of stopping the cold wind from blowing in. It had been a long cold winter, the ground outside had been frozen for months, and very little food was available. Tonight two women, four children, and an old man were busy with the carcass of a horse—a saber-tooth cat-kill they had found by the stream. Although hyenas, dogs, and birds of prey had already stripped it bare of flesh, its skull, backbone, and long bones were intact, sealing the precious marrow inside. Banging two rocks together, one of the women produced sharp flakes with which to cut the ligaments that held the horse's individual backbones in place. Another twisted and turned its head and spinal column, trying to detach the skull and separate the backbones to get at the marrow inside. Once detached, the skull was placed base-up at the edge of the fire, next to the long bones. One by one, the long bones were pulled out of the fire and with a large hammerstone cracked open. Their marrow was scooped out into a skull cap that served as a receptacle. When the long bones were finally emptied, the skull cap brimming with marrow was placed in the fire next to the horse's skull, and its contents stirred. In time, the group took turns slurping the marrow and eating from the horse's brain. The meal had to suffice for the night, for there was nothing else to eat.

At daybreak all that could be heard was the rushing of the wind through the barren branches of the trees growing by the stream. At first light, before sun up, the group left the cave to forage. Too weak to move, the old man remained by the dying fire at the cave's entrance. Three of the children ran off to dismantle a deserted beaver dam for firewood. At the beginning of winter the group had caught and eaten two of the beavers, and the remaining animals had moved away. The oldest child, a boy, went off by himself, following week-old deer trails along the stream, hoping to see some fresh tracks. The women had set off in the direction of three vultures they saw circling in the sky, hoping to find a carcass with more flesh than the horse they had found the day before.

When the children returned to the cave with the firewood, one of them noticed the old man was not moving. They stoked some embers covered in ashes, resuscitating a flame. In no time, the fire was ablaze. After warming up, they ran off to the stream to overturn rocks, hoping to find something to eat. By dusk the group was back in the cave. The cold and partly butchered corpse of the old man anchored the skins hung at the entrance, preventing them from flapping in the wind. Dead of hunger and old age, the old man's body had been partly consumed by the living, out of desperation.

HOMO PEKINENSIS

Skull, Teeth, and Diet

Homo pekinensis skull and teeth are numerous and thus fairly well known. They represent between thirty-two and forty individuals, with approximately one-third being children and two-thirds adults and adolescents. The *H. pekinensis* face is vertically set, but its upper and lower jaws jut out considerably. The cheek bones are very slender and notched where they meet the face. They attach high on the upper jaw at a level just in front of the wisdom tooth (third molar). There is a continuous bar of bone above the eyes, forming a brow ridge. The brow ridge curls downwards at either side of the eye sockets and is depressed and narrower at its midline. Along its length, it is guttered where it meets the braincase.

The *Homo pekinensis* braincase is low and long, narrowing in width behind the eye sockets. Braincase width from side to side decreases considerably with decreasing distance above the ear. The mean braincase volume for six individuals is 1059 cc, about two-thirds that of modern humans. Midline along the top of the braincase the bone is thickened, forming a keel. The skull base has more of a backward inclination than in modern humans or in other *Homo* fossils. Overall the bone that forms the wall of the skull case is characteristically thick.

The set of the *Homo pekinensis* teeth circumscribes a parabola, with exponential decrease towards the front in the width between corresponding teeth on opposing tooth rows. The molars wear flat. On average, molars and premolars are larger in size than in modern humans, with the exception of the last molar, which is reduced in size. Relative to the cheek teeth the front teeth are relatively smaller than in modern humans. Teeth appear to show greater size differences between males and females than they do in modern humans, approximating tooth size differences seen in male and female orangutans. Although the skull and skeleton also appear to shows size differences between males and females, these differences are not as marked as in the teeth.

There is no direct evidence as to the diet of *Homo pekinensis*, and with nearly all of the original teeth lost there is unlikely to ever be any. The flat and heavy wear on the teeth indicates a diet of herbs, seeds, nuts, roots, and tubers. A seasonal climate suggests meat-eating during times when food was scarce, and consumption of fruits and berries when these were in season.

Skeleton, Gait, and Posture

Aside from skull and teeth, all that is known of *Homo pekinensis* are seven thigh bone shafts, three arm bone shafts, a collar bone shaft, a shin bone fragment, and a wrist bone. The wrist bone is the only bone that preserves joint surfaces. It differs from modern humans in being broader and having joint surfaces with different curvatures and orientations, the significance of which has yet to be explored. The limb bones all have a thicker bone cortex than in modern humans. The thigh bone is more strongly flattened from front to back, with different muscular markings and shaft curvature than seen in modern humans. The angle between the

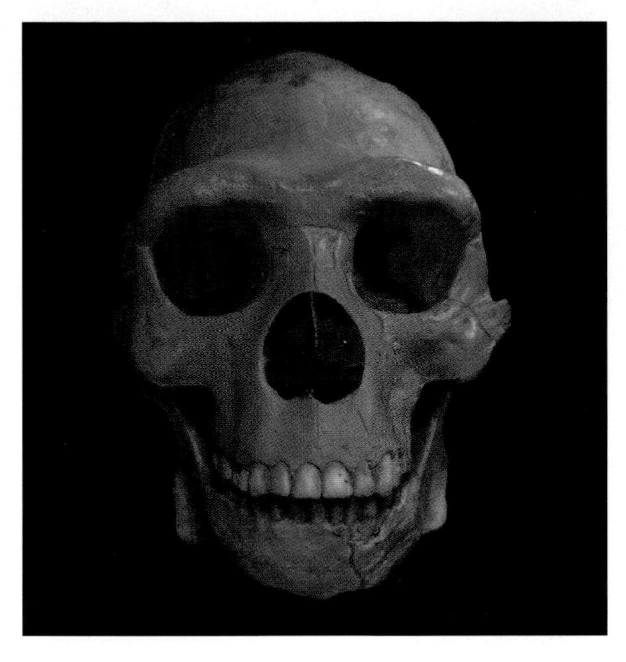

Composite reconstruction of skull of adult male *Homo pekinensis*, based on the original casts of the XII braincase, portions of mandibles GI and GII, and facial portions from skulls VI, X, and XIV from Zhoukoudian, China.

thigh bone neck and shaft is slightly lower than that of modern humans. As reconstructed by Franz Weidenreich, the most complete thigh bone shaft (Femur IV) is missing approximately 30 percent of its length and is relatively shorter and stouter than that of modern humans. This reconstruction is likely to be an overestimate, and the bone may even be shorter and thus stouter. The arm bone shows muscular markings not common in modern humans. Reconstruction of the most complete arm bone shaft (II) shows that 40 percent or more of its length is estimated. Its apparent slenderness may be simply due to reconstruction error. Arm-bone to thigh-bone length ratios of 0.80, halfway between modern humans and Lucy (A.L. 288-1), must be taken with a grain of salt considering how much of the long bone length is estimated, and the fact that thigh bone and arm bone are not from the same individual. The collar bone shaft is unremarkable and does not clearly differ from modern humans.

Based on femur length and assuming modern human proportions, Femur IV would have belonged to an individual with a stature of 156 cm, and Femur I to an individual 10 cm shorter. Although limb bone and wrist joint differences to modern humans may suggest differences in gait and posture, without joint surfaces or accurate long bone lengths its behaviors cannot be precisely reconstructed.

Fossil Sites and Possible Range

Homo pekinensis is known from a large collapsed limestone cave, which formed on the northern slope of Longgu-Shan (Dragon Bone) Hill, near the village of Zhoukoudian, 42 km southwest of Beijing, China. Because this cave was the first of two caves to be excavated and occurs on the same hill below the other younger cave, it is referred to as the "Lower Cave" or "Locality One." While open, the Lower Cave served as a shelter for *H. pekinensis*. Over time the cave filled up with sediment at depths of up to 40 m, and its roof collapsed. Excavators searching for *H. pekinensis* have divided the sediments that fill the Lower Cave into seventeen layers, with 1 being the most superficial and 17 the deepest. Layers 1-13 have produced either *H. pekinensis* fossils or evidence of human presence. Presence of *H. pekinensis* in the uppermost layers suggests humans were using the cave up until its collapse.

Considering possible overlap in age of deposits, *Homo pekinensis* may also be present at Yuanmou Hill on the banks of the most southern reaches of the Yangtze (Yellow) River, at Jianshi (Gao Ping) Cave, Hubei, or at Tham Khuyen Cave in Vietnam just south of the Chinese border. These fossils, however, are too incomplete to assign with any certainty to *H. pekinensis*. Hexian Cave, just west of the Yangtze delta, has deposits contemporaneous with Zhoukoudian, but yielded a skull that differs considerably from that of *H. pekinensis,* underscoring the uncertainty of classifying incomplete singular fossils.

Age

Paleomagnetic analysis shows that during the time layers 1-13 were deposited, Earth's magnetic pole was positioned as it is today. Below layer 13, the

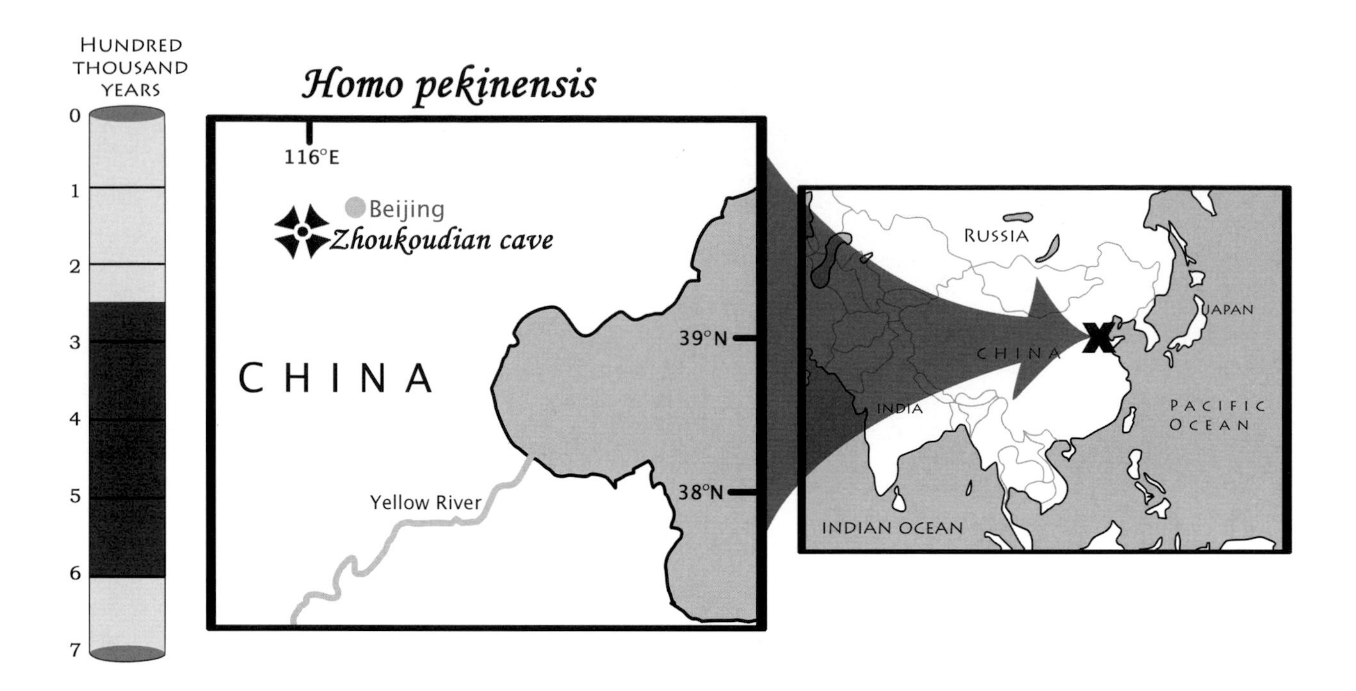

sediments show a reversal in Earth's magnetic pole that is known to have occurred approximately seven hundred and eighty thousand years ago. This date provides a lower age limit for all *Homo pekinensis* fossils. A number of absolute dating methods have been attempted for layers 1-13, including some directly on bone and teeth, but all have considerable error. The most recent dates reported for some of the layers are: 220,000 to 249,000 years (layer 2); 249,000 to 350,000 years (layer 3); 292,000 to 320,000 years (layer 4); 368,000 years (layer 6); 385,000 to 396,000 years (layer 7); 423,000 years (layers 8-9); 462,000 to 505,000 years (layer 10); and 578,000 to 585,000 years (layer 11). Most *H. pekinensis* fossils come from layers 8-9, and thus have an age of about four hundred and twenty thousand years. Recent absolute dates based on cave flowstones suggest the majority of fossils may be as much as six hundred thousand years old.

Tools

Layers 1-13 have yielded over 100,000 stone artifacts, consisting mostly of flakes and simple chopping tools. Reminiscent of Oldowan tools, very few of these tools show flaking on more than one side. For the most part, tools found in the earliest (deepest) layers are large and were made of sandstone or other soft stones. Over time the tools became smaller and were made largely of quartz. The smallest and most finely worked tools are found in the most recent layers. Here sandstones or other coarse materials are no longer used, and nearly 30 percent of the tools are made of fine grain flint.

Early claims that Zhoukoudian shows evidence that *Homo pekinensis* used fire in layers 4-10 have recently been questioned. Although some bones appear to be blackened, there is no ash and thus no direct evidence of domestic fires. Earlier claims of cannibalism based on broken skull bases and long bones have also been questioned. The broken bones are currently interpreted as the work of hyenas. None of this, however, is certain.

Animals and Habitats

Mammals found at Zhoukoudian are typical of temperate climates and modern in appearance. A beaver (*Trogonotherium*), the woolly rhinoceros (*Coelodonta*), and a saber-tooth cat (*Machairodus*) are the only mammals that belong to extinct genera. Living genera include shrews, moles, hedgehogs, bats, pikas, squirrels, groundhogs (marmots), porcupines, voles, field mice, harvest mice, bamboo rats, gerbils, dogs, hyenas, Asiatic black bears, cave bears, wolverines, cheetahs, forest elephants, one-horned rhinoceroses, horses, boars, deer, elks, Asian water buffaloes, cattle, and macaques. Recently, orangutans have also been reported to occur in the same deposits with *Homo pekinensis*. Considering a temperate climate, if these isolated teeth do indeed represent a great ape, it is unlikely that it could be the same species or even genus as the modern tropic-living orangutan.

Despite being more than a million years younger and 8000 km further to the east, Zhoukoudian has many of the same animals found at Dmanisi, Georgia (*see Homo georgicus,* Animals and Habitats), reflecting similar latitudes and temperate climates with cold winters. Other alleged *Homo* sites in mainland Asia at more southern latitudes yield prehistoric elephants (*Stegodon*), the giant panda, and *Gigantopithecus*. The latter is an extinct great ape relative that, based on its large dental size, is claimed to be the largest primate that ever lived.

Climate

Beijing, 42 km northeast of Zhoukoudian, reports a typical temperate climate. Reflecting extremes, mean monthly temperatures throughout the year may vary by as much as 30° C (83° F). The winter months (December-February) all record monthly mean temperatures below freezing, and freezing temperatures may occur October-April. The summer months (June-August) all record monthly means above 24 °C (75° F). In August, temperatures may reach a high of 41.7° C (107° F). Mean annual rainfall is 638 mm, with June-August receiving most of the rainfall (472 mm, 74 percent). The winter months (November-March) receive only 20 mm of precipitation, usually as snow. Although during the time *Homo pekinensis* occupied Zhoukoudian at least three glacial periods were recorded, it is unclear if during these glacial periods temperatures were much lower than they are today.

The associated animals suggest winter temperatures probably did not drop appreciably during glacial periods. Glacial periods, however, may have been associated with colder summers and wetter conditions. The temperatures, even during interglacials, would have been too cold in winter for *Homo pekinensis* to live without fire, clothing, or shelter.

Classification

Given some shared skull features associated with a relatively small braincase, and an average braincase volume that is only 100 cc greater in *Homo pekinensis* than in *Homo erectus* from Sambungmacan, Trinil, and Sangiran, Java, some paleoanthropologists have assumed that all these hominids represent the same species. *Homo erectus* and *H. pekinensis* skulls, teeth, and especially thigh bone differences do not justify such an assumption. Until more material is uncovered, it is more prudent to keep the two as separate species.

Homo pekinensis–like thigh bones from Olduvai and Koobi Fora suggest that it or a closely related species may have also occurred in Africa. At the moment, however, these fossils are too incomplete to decide whether this similarity is the result of a close relationship, or only of a similar function.

Historical Notes

The Geological Survey of China began excavations at the Zhoukoudian Lower Cave in 1921 under the direction of Otto Zdansky. He is credited with finding the first fossils of *Homo pekinensis* in the summer of 1926, an upper molar and lower premolar, which he recognized when cleaning fossils in a laboratory in Upsala, Sweden. In a 1927 *Bulletin of the Geological Survey, China*, Otto Zdansky reported these finds as belonging to an indeterminate *Homo* species. In a 1927 issue of *Paleontologica Sinica*, Davidson Black coined the name *Sinanthropus pekinensis* for the two teeth found by Zdansky and an additional lower molar found in October of that same year. Black, however, did not elaborate as to why the

three teeth were distinctive at the species or genus level. In a 1928 *Paleontologica Sinica,* Otto Zdansky described fossil mammals he collected from the Lower Cave. This was followed by Franz Weidenreich's publications in *Paleontologica Sinica* on the *Sinanthropus pekinensis* lower jaw bones, teeth, limb bones, and skull in 1936, 1937, 1941, and 1943.

In November 1941 in response to Japanese occupation of China, all *Homo pekinensis* fossils found up to that date (with the exception of the two teeth found by Zdansky) were packed and arranged to be taken from Beijing to the United States. The shipment with all the fossils was lost and never recovered. Weidenreich, however, had made accurate casts of all the fossils, and these are presently housed in the American Museum of Natural History. Soon after its descriptions, most paleoanthropologists began referring to this species as *Homo erectus* (see *Homo erectus*).

Post World War II finds of *Homo pekinensis* were made in 1951, 1955, and 1966 and include a shin bone fragment, an incomplete lower jaw bone, and braincase fragments. The braincase fragments correspond to fossil remains of an individual that was found prior to 1941 and lost in shipment. Original *H. pekinensis* fossils are housed in the Department of Geology and Paleontology University of Upsala Sweden, and in the Institute of Vertebrate Paleontology and Paleoanthropology Beijing, China.

The species name "*pekinensis*" is from the old spelling (Peking) for Beijing. "*Sinanthropus*" is from the prefix "Sino," meaning China (from the Greek word "Sinai") and the suffix "anthropus," meaning man.

Days of hunger compel Peking Man to hunt for his food deep in the forests of China some three hundred thousand years ago. Stalking his prey, he hides behind a tree and waits for the right moment to move in for the kill.

AN ISLAND SUNSET FOR THE LITTLE PEOPLE

He was only a three year old when he first set eyes on the giant men. What struck him most was not just how big they were, but the size of their heads and how long and thin their legs were. Those who saw the giant men when they first stepped on the island told stories that they arrived on hollow tree-trunks from across the big water like giant fish. That was many years ago when there were still many elephants and giant lizards to hunt. Now that he was a man, the only elephants that remained were in the forested hills where he lived with his family. The giant men did not know there were still elephants here, or they would have eaten them all. Today he was worried. He had picked up the trail of one of these giants close to his home. There was no mistaking the long thin foot prints they left behind. He tried to determine from what direction the prints were coming, but lost the trail in a rocky clearing. He climbed up a tree to get a better look and was able to spy a giant miles away walking through a clearing in the valley.

It was late afternoon by the time he reached the spot where he had seen the giant. Along the forest edge he noticed two snare traps the giant had set. One of them had a live giant rat in its noose. He hit the rat over the head with a stick until it was dead, tied the tail and forelimb of the rat together, and slung it over his shoulder. He followed the giant's trail along the edge of the forest. After walking for some time, he saw the thatched hut the giant had built along the forest's edge. Crouching down, he crept up on the hut to get a better look.

The giant kept live animals around his home. One he had seen before. It was always found together with giants. It did not like people and would bark at them, occasionally chasing and biting them. Long ago his brother had been chased down by two of them. They bit him and tried to eat him, but he was able to escape. The other animal was a fat one. It looked like a small elephant with short legs, short tusks, and small pointed ears, but without a trunk. It snorted and rooted in the dirt. Although he had never eaten one, he heard they had juicy white meat and tasted very good. The giant had made an enclosure out of branches so that it could not escape.

He heard voices and saw a woman, not a giant one but a normal one like him, breast feeding a baby. Although the giants had brought many animals with them, they had brought few women. In fact, he had never seen a giant woman, although he heard they existed. All the women he had ever seen with giants were stolen from his people. He thought about his wives and children at home. The thought that one of his wives or, worse, one of his daughters could be stolen by a giant disgusted him. He did not like giants. They were bad. They stole women and ate all the animals.

It was getting late and it had started to rain. He turned around and headed back home. Just before dusk, there was a torrential downpour. He ran to a nearby ledge to wait it out. When darkness came, the thick cloud cover screened out all the light from the moon and stars. He could not even see his hand in front of his face and was forced to spend the night below the ledge. The following morning, he was up at dawn anxious to return home. A bad dream about the giants made him fear for the safety of his family. Overturning rocks, he found some grubs which he quickly ate. Crossing a stream swollen by last night's downpour, he saw a large fish stranded in an isolated pool of water. He repeatedly tried to scoop the fish out with his hand, but it evaded him every time. Aided by a stone, he dug a hole in the side of the pool and let the water drain out catching the fish by the gills. By mid-morning he was back. His wives and children were there to greet him. The fish and rat he had brought back would make a fine meal.

He sat at the entrance of the cave he called home, retouching and sharpening his spears. Tomorrow he would go out with his sons to hunt for elephant. If he didn't eat the elephants, the giants would eat them all. Looking towards the sunset, he wondered how much longer it would be until the giants moved into the forest and took over the entire island.

HOMO FLORESIENSIS

Skull, Teeth, and Diet

The skull and jaw bone associated with a skeleton (LB1) and a separate jaw bone (LB6/1) are all that is known of the *Homo floresiensis* skull. Although the LB1 skull is complete, its nose, left eye socket, and brow ridge are damaged, and the face, upper and lower jaw, and inside of the braincase are slightly distorted. Nevertheless, this skull affords a wealth of detail that serves to distinguish it from modern humans.

The skull is very small with a very small vertically set face. The skull case is thick-walled and fairly globular in shape, with a slight side-to-side constriction behind the eye sockets. The forehead is relatively high, but overall the skull vault is low. A cast of the inside of the skull case shows a small brain, markedly flattened from top to bottom. Direct measurements on the original fossil yielded a brain volume of 380 cc. This volume is slightly less than the 417 cc arrived at through three-dimensional computer reconstructions. Both values are close to the average brain volume of living chimpanzees and below that of gorillas.

The eye sockets are circular in outline, and each has a small, arched brow ridge above. In contrast to modern humans, the bone separating the nose opening from the mouth is very narrow. On the palate just behind the front teeth, there is an opening for nerves traveling between the nose and the roof of the mouth, which is uncommonly large when compared to that of humans. As in humans, the canines are small, show only tip wear, and do not protrude past the tooth row. The first lower premolar, however, is peculiarly elongated and has double roots. The upper premolars have a correspondingly unique shape. As in most *Homo* species, the lower first molar is the largest molar and the third molar the smallest. Both upper and lower front tooth rows are relatively narrow and the front teeth small.

No studies of the *Homo floresiensis* diet have yet been undertaken. Tooth shape suggests that it and its ancestors ate a wide range of foods, which included vertebrates, insects, fruits, seeds, tubers, and herbs. *Stegodon* remains associated with tools suggest that it may have hunted and eaten this prehistoric pygmy elephant. The presence of large lizards, turtles, frogs, fish, and rats in the cave deposits suggests that these animals (many of which are not normally cave dwellers) may have also been eaten by *H. floresiensis*. Given its small body size, insects may have provided a greater percentage of the diet than in much larger extinct hominids.

Skeleton, Gait, and Posture

Eight out of a minimum of nine *Homo floresiensis* individuals found at Liang Bua are represented by fossils other than the skull and teeth. Nevertheless, most of what is known of this hominid comes from the LB1 skeleton, which is the most complete *H. floresiensis* individual so far reported. The LB1 skeleton includes a nearly complete left hip bone, and all of the limb long bones with the exception of one of the forearm bones. Although all the long bones are somewhat damaged, they are

Reconstruction of the damaged LB1 *Homo floresiensis* "Hobbit" skull from the Javanese island of Flores.

complete enough to yield relatively accurate length estimates. *Homo floresiensis* arm-bone (243 mm) to thigh-bone (280 mm) length ratio is 0.868. This value is outside of the modern human range, and also outside of the range of all living African apes. It is slightly greater (by 0.02) than in *Australopithecus afarensis* (Lucy, A.L. 288-1), and very close to the baboon average. All *H. floresiensis* limb long bones have thickened bone and are much stouter with greater mid-shaft-diameter to length ratios than in modern humans. Hip bone, arm bone, and foot bone muscle attachment areas and joint surfaces show different orientations than those commonly seen in modern humans. Relative to hip bone dimensions and thigh bone mid-shaft cross-section, the hip joint is considerably smaller than that of modern humans. Forearm and shin-bone curvatures also differ from the common human condition.

Based on thigh bone length, the LB1 individual's stature is estimated at 106 cm, and its weight at 16 kg to 28.7 kg. The LB8 individual, with a shin bone 1.9 cm shorter than LB1, would have a somewhat shorter stature (97.5 cm) and possibly a lower body weight. Because these estimates are based on modern human pygmies, from whom *Homo floresiensis* differs considerably, their accuracy is questionable. For instance, weight estimates using thigh bone mid-shaft cross-sectional area yield a considerably greater weight (36 kg).

The marked differences in the *Homo floresiensis* skeleton, when compared to modern humans, must have a functional explanation. Given its small body size and an arm-bone to thigh-bone length ratio that is baboon-like, *H. floresiensis* was likely more adept in trees than modern humans. Its stout long-bone shafts and arm-bone to thigh-bone length ratios are more in accord with a four-legged walker that orients the long axes of its limb bones oblique to the force of gravity. With shoulder blades, backbones, and hand and foot bones yet to be described, it is premature to jump to conclusions as to *H. floresiensis* behaviors.

Fossil Sites and Possible Range

Homo floresiensis is known only from Liang Bua Cave, a limestone cave 500 m above sea level on the Indonesian Island of Flores. The cave is approximately five degrees east (560 km) from the eastern end of Java, and just two degrees east of the Wallace-Huxley Line. This line marks the edge of the Indonesian continental shelf. During Pleistocene glaciations, low ocean levels (due to a larger percentage of Earth's water existing as ice) exposed the shelf, providing land mammals from continental southeast Asia access to those Indonesian Islands west of the line (mainly Sumatra, Java, Borneo, and Bali). East of the Wallace-Huxley Line and off of the continental shelf, the ocean level is deep enough that it acts as a barrier to animal movement from west to east, even during low ocean levels.

Liang Bua Cave originally formed as a result of underground water dissolving the surrounding limestone. In time, the Wae Racang River invaded

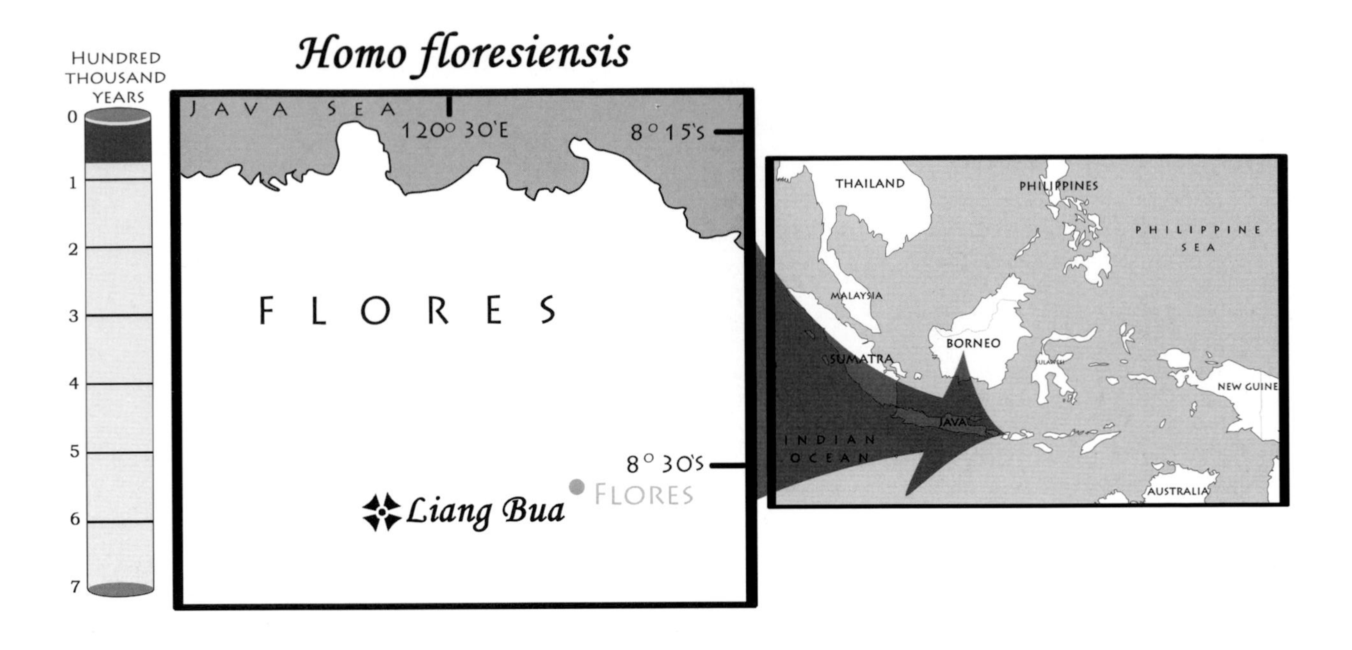

the cave, forming its northern opening. The river now passes 30 m below and 200 m away from the cave opening. The cave floor is covered in sediments that in some sections of the cave surpass 11 m in depth. At the base of these sediments are conglomerates consisting of water-rolled rocks of varying sizes. The base conglomerates record the river's past course through the cave. Above the conglomerates, the overlying sediment is mostly clay inter-lain with flowstone and washed-in volcanic ash. Both the flowstone and volcanic ash layers may be used to date the fossil remains. In one section of the cave, layers above volcanic ash contain modern human remains recording the last 10,000 years. The holotype specimen skeleton (LB1) was found in fine clay below a thick volcanic ash layer at a depth of 5.9 m against the cave's east wall, not far from its entrance.

Age

A thick volcanic ash layer overlying all but one *Homo floresiensis* fossil yielded a radiometric date of 13,100 years. A child's radius, found 40 cm above this level, was estimated to be 12,000 years old. Radioisotope dating of carbon residues associated with the LB1 skeleton yielded a date of 18,000 years. Two other absolute dating methods on sediments associated with the LB1 skeleton provided a maximum and minimum date that further verified the carbon isotope date. In another section of the cave, absolute dating of a flowstone layer overlying an isolated *H. floresiensis* lower premolar (LB2) and forearm bone (LB3), and of a *Stegodon* tooth just below the premolar, yielded dates of 38,000 and 74,000 years. Bracketed between these dates, the LB2 and LB3 fossils are older than the LB1 skeleton. *Homo floresiensis* remains at Liang Bua, therefore, appear to span a maximum time interval between 12,000 and 74,000 years ago. A flowstone, overlying the base conglomerate in the back of the cave, yielded a radioisotope date of 102,000 years, which is a minimum age for the oldest sediments at Liang Bua.

Stone tools and other evidence of human occupation on Flores Island may date back to 840,000 years. If, as seems likely, *Homo floresiensis* is an authentic species, it will probably be associated with at least the last 400,000 years of this evidence. Species as distinct as *H. floresiensis* do not appear suddenly. In this case, it should be found in other Flores fossil bearing deposits extending further back in time than at Liang Bua.

Tools

Stone tools occur throughout the Liang Bua Cave deposits in which *Homo floresiensis* is found. Tools are made of volcanic rocks or chert (a rock chemically precipitated in ocean water) and flaked-off from a prepared core. Most tools found are simple flakes that have been modified on both sides. Associated with *Stegodon* fossils, however, there is also a "big-game" tool kit that includes points, blades, punches, and micro-blades. A large chopper reflecting a simpler tool technology, found within the base conglomerate and predating all the *H. floresiensis* fossils, may also belong to this species.

Differences Between Males and Females

Based on comparisons to modern human hip bones, the LB1 skeleton describers believed it belonged to a female. The more robust LB1 vs. LB6 lower jaw suggests that LB1 is more likely a male. With so few remains, however, such sex determinations are far from certain.

Animals and Habitats

The animals found in the Liang Bua Cave have not been fully reported. Only rodents, bats, some large lizards, including the Komodo dragon, turtles, frogs, and fish have been mentioned. There is also a pygmy *Stegodon* (a prehistoric relative of elephants). Because Flores is east of the Wallace-Huxley Line, very few flightless mammals are known to naturally occur on the island. Many mammals on Flores today (e.g., macaque monkeys, porcupines, rusa deer, pigs, palm civets, mongooses, dogs, house rats, and mice), and in the more recent deposits at Liang Bua and other caves, have most likely been brought to Flores by modern humans.

Skeletal remains suggest modern humans arrived on Flores between 35,000 and 55,000 years ago and for a minimum of 23,000 years would have coexisted with *Homo floresiensis*. Undoubtedly, attempts at reproduction occurred, and it is possible that such crosses would have produced fertile offspring. It is unknown whether *H. floresiensis* was absorbed into the modern human population or became extinct due to competition or disease linked to modern humans.

With the exception of the two *Stegodon* species (one most likely predating the other) and *Homo floresiensis*, the only flightless mammals believed to be native to Flores are a shrew and six closely related rat species. Of the latter, three are extinct (*Spelaeomys*, the Flores cave rat, *Hooijeromys*, Hooijer's rat, and *Papagomys theodoverhoeveni*, Verhoeven's giant Flores rat). The giant Flores rat (*Papagomys armandvillei*), the Komodo rat (*Komodomys*), and the long-nosed Flores rat are still found in Flores, and the Komodo rat is also found in the Komodo islands of Rintja and Padar. So far only Hooijer's rat is known from Pleistocene deposits contemporaneous with *H. floresiensis*, although it is likely that some of the other native Flores rats extend into the Pleistocene. Weighing in excess of one kilogram, *Papagomys* and the extinct giant rats would have been eagerly sought by *H. floresiensis* as food.

Giant rats and the pygmy *Stegodon* presence on Flores support the likelihood that *Homo floresiensis* is a *bona fide* species and not a dwarf microcephalic modern human, as has been suggested by some. Over long periods of time on small islands with few competitors or predators, small mammals tend to become large, and large mammals tend to become small. Island size strongly limits space and resources for large mammals, but not for small ones. On a small island, large animals will decrease body size to increase population number per given area. This leads to greater population numbers and variability, decreasing the odds of becoming extinct. Additionally, in mammals that use their size as a defense from predators, predator absence relaxes selective pressures to maintain large size. On small islands, without competitors or predators, there is no check that prevents small mammals from increasing in size in order to out-compete each other for resources. Size increases also may make resources available to smaller sized mammals that were otherwise unavailable. If both rats and the pygmy *Stegodon* underwent changes in body size in response to an isolated life in Flores, it makes sense that *H. floresiensis* ancestors would have also.

Climate

Considering that *Homo floresiensis* survived until 12,000 years ago, the climate during its existence would not have been much different than it is now. Kupang, a city on Timor Island just west of Flores and further south than the Liang Bua Cave, records temperatures that are 1.5° C (3° F) colder than the coastal city of Surabaya, Java (see *Homo erectus* Climate), but shows more or less the same temperature pattern. July records the coldest (26° F) and September through November (28° C) the warmest mean monthly temperatures. Mean annual rainfall in Kupang (1410 mm) is slightly less than in Surabaya; however, Kupang's more southern latitude results in greater seasonal differences. December through March (1100 mm) is a bit wetter, and May through October is drier (70 mm). At latitudes halfway between Kupang and Surabaya, Liang Bua should be approximately 0.8° C (1.4° F) warmer than Kupang, with similar mean annual rainfall and slightly less seasonal differences in rainfall pattern. *Homo floresiensis* would not have needed clothing. Fruits and plants in the diet would have been available year round.

Classification

If *Homo floresiensis* represents a microcephalic, it certainly does not represent a modern human one. It exhibits too many unique skeletal features that distinguish it from modern humans and also modern human microcephalics. With a small brain size close to the chimpanzee average and a skeleton that is claimed to be *Australopithecus*-like but has yet to be described in detail, it is not clear why *H. floresiensis* is assigned to *Homo*. Its small brain size, unique premolars, unique foot bones, and unique palate, teeth, and face all suggest a new genus may be more appropriate. At this point, however, it is best to wait for a complete description and hopefully more fossils. Its relatively recent age makes it likely that remains preserving DNA will be found. This should help clarify its classification.

Historical Notes

In 1950, Theodor Verhoeven visited the Liang Bua Cave when it was used by nearby residents as a school. In 1965, Verhoeven excavated the cave and found recent human graves and fossils from one of the Flores endemic rats, *Paulamys*. Peter Brown and colleagues found the *Homo floresiensis* skeleton in September 2003. In an October 2004 *Nature*, they made the newly found skeleton the holotype specimen of a new species, *Homo floresiensis*. Persuaded by its small size and brain volume, the group originally pondered referring the species to *Australopithecus*, but opted for the more conservative *Homo*. A month later in press releases picked up by *Science* and *Nature*, Macjie Hennenberg and Robert D. Martin claimed the fossil is a modern human microcephalic. In October 2005, the Liang Bua group reported on fossils of eight more individuals, including a child. The fossils are housed in the Center for Archaeology in Jakarta, Indonesia. The species name "*floresiensis*" refers to the island to which this fossil appears to be endemic.

A survivor of over one and a half million years of isolated evolution, this female *Homo floresiensis* is one of the last of her kind on the tropical island of Flores, Indonesia, some twelve thousand years ago.

AFRICA, EUROPE, AND THEN THE WORLD

Homo antecessor • *Homo rhodesiensis* • *Homo heidelbergensis* • *Homo neanderthalensis*

Somewhere between Africa and Europe within the last million years, a new lineage of human emerges, one that is more dependent on technology and is at home as much in temperate as in tropical areas. The origin of modern humans lies within this lineage.

THE ULTIMATE COMPETITOR

Every year at the end of autumn when the food supply dwindled, fierce competition between humans and hyenas ensued. This competition was not just limited to access to animal carcasses, but included humans killing hyenas and hyenas eating humans. Humans both hated and feared hyenas.

Recently a hyena had wandered into camp in the middle of the night. Smelling the breath of its victim, it clamped its jaws around her face and dragged her off. Her younger sister claimed to have seen it happen and was hysterical. She alarmed the rest of the camp. A group set out in the night to look for the girl and the hyena. One of the boys believed the hyenas lived in a nearby cave. When the group went to the cave, they failed to find the girl or any hyenas.

They searched until daybreak. In the daylight, the signs of where the hyenas dragged the girl were obvious up to a point. After that, the trail divided and seemed to include several hyenas. It was probably here that the girl's body had been dismembered, each hyena carrying away a part, but there was no blood or pieces of the girl to be found anywhere. It was the second day in a row hyenas had entered the camp. The day before, the hyenas had taken the carcass of a deer. The group had now stashed all of their uneaten food high in a tree where the hyenas would not be able to reach it.

Distraught at the loss of the girl, the group slept uneasily the following night. Their uneasiness was heightened by the whoop calls of hyenas, which seemed to get closer and closer as the night progressed. At one point the hyenas were so close, the low pitched growl that precedes the hyena's whoop call could be heard. When this occurred, those who heard it stood up and threw stones in the direction of the growl. They flushed the hyena out of a bush, and pelted it until it dropped to the ground. The hyena, however, stood up and scurried away, with the humans giving chase. Running after it, they managed to corner another hyena against a shallow rocky crevice and beat it to death with sticks and stones. Once it was dead, they pulled the animal apart and brought its body parts back to camp.

For the next two weeks hyena calls were heard every night close to camp. The killing of the hyena, however, had somehow eased the tension and once again the group slept soundly. One day, however, the failure of two of its members to return from a day's outing raised the group's suspicions that hyenas were responsible.

On the opposite side of the mountain from camp, looking to catch porcupines in two remote caves, a young man from the group discovered a cave with the broken skulls and burned bones of humans. Believing he had finally found the hyena den, he staked out the cave entrance to await their return, hoping to kill at least one of them. Just before nightfall the young man witnessed two men, carrying a partial human carcass, walk into the cave. Pulling out some embers they had stashed at the rear of the cave, they started a fire. With sharp stone flakes in their hands, they butchered the remains, and sat down to enjoy their meal. The young man did not confront the cannibals. Instead he ran back to camp to tell the others that the hyenas eating them had magical powers, and could turn into men. In reality, with the absence of large carnivores and primates, competition between humans had become more intense.

HOMO ANTECESSOR

Skull, Teeth, and Diet

There are no complete skulls of *Homo antecessor*. What is known of the skull is known from fourteen skull fragments and two incomplete lower jaw bones (ATD 6-5 and ATD 6-96), belonging to a minimum of eight and a maximum of fourteen individuals. The two most complete remains are a child's upper jaw and left cheek bone (ATD 6-69) and an adult's forehead bone, preserving the sinus airspaces, brow ridge, and upper eye socket border (ATD 6-15). There is also the left front quarter of a child's upper jaw (ATD 6-14) and a number of cheek bone fragments (ATD 84, ATD58, ATD 38, and ADT 19). Based on modern human age and tooth eruption relationships, the upper jaws ATD 6-69 and ATD 6-14 belong to a twelve-year-old child and a five- to seven-year-old child. The remaining fragments are from the skull base. Some of these preserve part of the middle and inner ear (ATD 6-18), the backbone (ATD6-77), and jaw joint surfaces (ATD6-17a).

Aside from the teeth in the ATD 6-69 (front tooth, canine, and seven cheek teeth) and ATD 6-14 (milk canine and milk molar) upper jaws, and in the ATD 6-5 (three molars) and ATD 6-96 (two premolars and three molars) lower jaws, there are an additional sixteen adult *Homo antecessor* teeth known. Two of these, a canine and premolar, are from a very incomplete upper jaw (ATD 6-13); the remaining fourteen teeth are isolated finds. Eleven of these together with ATD 6-5 molars and an ATD 6-13 canine and premolar are claimed to belong to the same twelve- to fourteen-year-old child. The other three are front teeth, which are believed to come from three separate individuals. The teeth suggest there are at least seven individuals represented in the deposits.

According to its describers, the *Homo antecessor* skull shows some of the following defining features: 1) a fully modern mid-face with a cheek bone that attaches high on the upper jaw, and a cheekbone surface that faces front and slightly downwards; 2) a single brow ridge molded in an arch above each eye; 3) a brain volume that is above 1000 cc; 4) a jaw that is not as thick as that of *Homo ergaster* or *Homo habilis*; 5) lower front teeth that are wider from front to back than *H. ergaster* and *H. habilis*; 6) a lower canine that is narrow from front to back; 7) premolars that are wide from side to side; 8) lower premolars with two roots and three pulp canals each and upper premolars with two well separated roots; and 9) cheek teeth with wrinkled enamel that are relatively smaller than *H. habilis*, a first molar that is smaller than the second molar, and wisdom teeth that are small relative to the other molars.

Despite belonging to a relatively young individual, the ATD 6-69 upper jaw and left cheek bone show a well formed spine for attachment of the nose septum, which scallops the border of the lower nose opening. Such strong development is otherwise shared only by *Homo heidelbergensis*, *Homo neanderthalensis*, and modern humans. Although *Homo antecessor* is claimed to have a braincase volume above 1000 cc, no specific braincase volume has been reported. This may prove difficult to determine given the incomplete nature of the skull fragments.

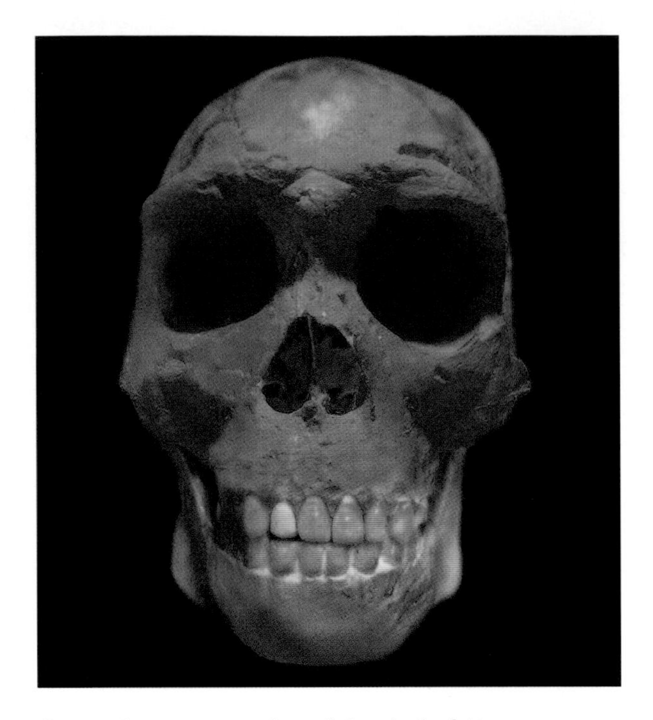

Composite reconstruction of the skull of *Homo antecessor* based on skull fragments from the Gran Dolina Cave, Sierra de Atapuerca, Spain; a mandible from the Ternifine Quarry near Mascara, Algeria; and a partial skull from Daka, Ethiopia. While the Spanish and Algerian fossils have been dated in the same 700,000- to 800,000-year-old span, the Ethiopian skull is over 200,000 years older and separated from the other two by upwards of over 3000 km and they may not represent the same species. This reconstructed skull is, therefore, the most conjectural of those pictured in this book.

Similarities in *Homo antecessor* teeth to those of *Homo heidelbergensis* suggest a similar diet. Their cheek tooth and lower jaw size dimensions indicate this diet probably included abrasive foods. *Homo antecessor* would probably have relied on herbs, seeds, tubers, and roots, but eaten nuts, fruits, and mushrooms when these were in season. Meat would have been most often consumed in winter when plant material was scarce.

Skeleton, Gait, and Posture

The *Homo antecessor* skeletal remains are fragmentary, with few preserved long bones with joint surfaces. From the upper limb there are three collar bones. Two of these are complete with joint surfaces, but only one is from an adult (ATD 6-50). The adult bone is long (161.5 mm) both absolutely and relative to its mid-shaft diameter when compared to that of modern humans. Together with a forward orientation of its shoulder blade joint surface and marked shaft curvature, its length suggests a wide chest from both side to side and front to back. Such a chest dimension is also suggested by the first and second rib (ATD 6-108 and ATD 6-79) curvature.

A forearm bone shaft missing only the joint ends (ATD 6-43) provides an accurate indication of forearm length (257 cm). It is much longer absolutely and relative to its mid-shaft diameter than those of Neanderthals, *Homo heidelbergensis*, and human Portuguese populations from the nineteenth and twentieth centuries. Assuming human body proportions based on means for modern human males and females, *Homo antecessor* height, as estimated from the forearm bone, would have been 172.5 cm.

Of the lower limb bones, there is only an upper thigh bone shaft. It is too incomplete to provide much information. Two vertebrae from the neck region (ATD 6-9 and ATD 6-51) and two kneecaps are smaller in size than average for modern humans, but otherwise similar in most respects. When compared to Neanderthals and *Homo heidelbergensis*, the knee caps are narrower from side to side, but, like Neanderthals and in contrast to modern humans, the joint surfaces corresponding to the thigh bone are sub-equal in size.

Two wrist bones (ATD 6-23 and ATD6-24), two incomplete hand bones (ATD 6-59 and ATD 6-26), and eight finger bones are in most respects similar to those seen in modern humans and not as stout as those of Neanderthals or *Homo heidelbergensis*. The same is true of eight toe bones. A complete foot bone (ATD 6-70+107; 77.7 mm in length) is longer than average for Neanderthals and close to the modern human average length. Assuming average European male and female body proportions, the individual belonging to this foot bone

Homo antecessor

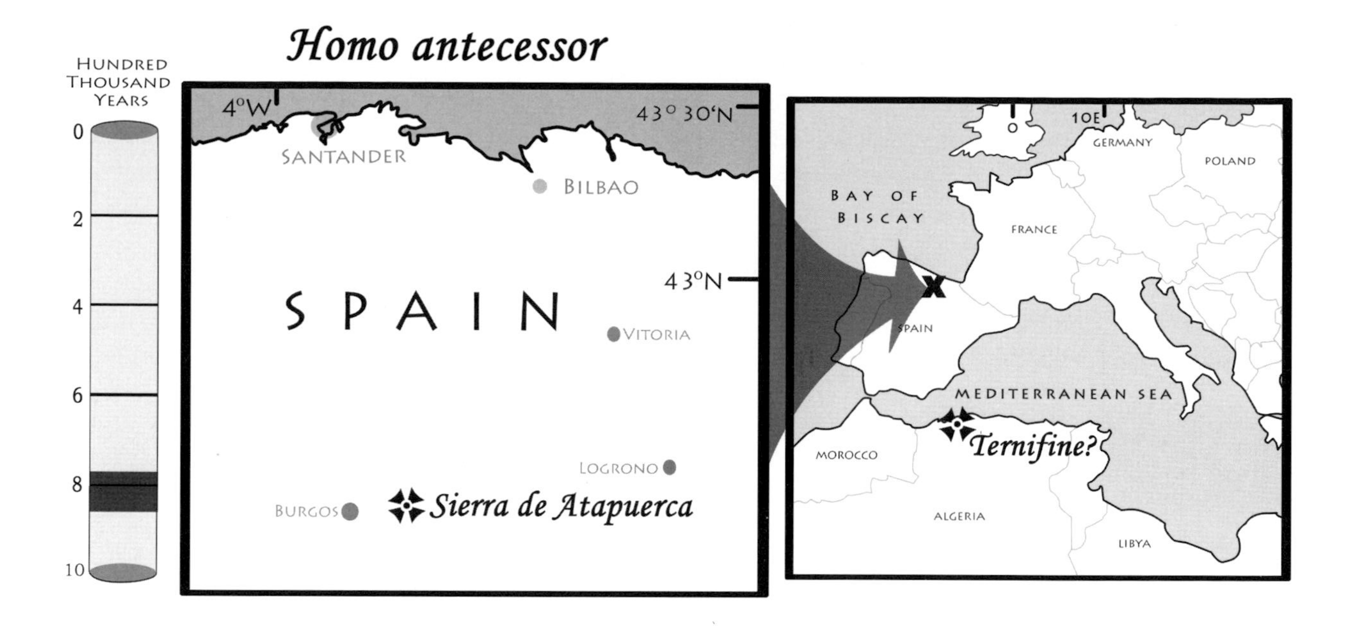

had a stature of 171 cm. Although body proportions and bone joint size and orientation, especially for the long bones, are not known, the strong similarities to humans in the hands and feet indicate *Homo antecessor* stood and moved around on two legs as do modern humans.

Fossil Sites and Possible Range

Homo antecessor is known from the aurora stratum (upper part of TD6) of the Gran Dolina (Great Sinkhole) Cave deposits in the Sierra de Atapuerca, 14 km east of Burgos, Spain. Gran Dolina is one of a series of collapsed limestone caves whose sediments were transected when a 1.5 km-long railroad trench was cut into the Sierra de Atapuerca between 1896 and 1901. The other cave deposits transected are Sima del Elefante (Elephant Shaft), El Penal (the Penitentiary), Cueva de los Zarpasos (Lash of the Claws Cave), Galeria (Gallery), and Tres Simas Boca Norte (Three Shafts North Opening). The latter three are part of the same cave system are all referred to the Galeria complex. The other fossil bearing cave in Sierra de Atapuerca, Cueva Mayor (Major Cave), containing the Sima de los Huesos deposit (*see Homo heidelbergensis*), was not transected by the railroad. Its opening can be found approximately one kilometer south-southeast of Gran Dolina.

All of the caves formed in 70 m-thick dolomite, which was laid down in shallow seas during the Cretaceous period (80 million to 100 million years ago) and later uplifted. The dolomite was exposed to water action around Late Miocene times (5.3 million to 7 million years ago), and the caves began to open sometime around one million two hundred thousand years ago. Gran Dolina was filled up with sediment when it collapsed between 200,000 and 300,000 years ago. Excavators divided the approximately eighteen meters-thick Gran Dolina Cave sediments into eleven layers, TD11-TD1 (the TD prefix stands for Trinchera (trench) Dolina), from most superficial to deepest. The TD11 layer contains rubble from the roof collapse, and is partly exposed on the land surface. Although human fossils have been reported only from Gran Dolina, Sima de los Huesos, and the Galeria complex, the other cave sediments, parts of which are contemporaneous with the hominid bearing deposits, also provide a glimpse into animal life and habitat in the Sierra de Atapuerca during the time in question.

It is possible that *Homo antecessor* is also known from Ternifine (Tighenif) Quarry, 20 km east of Mascara, Algeria. With three lower jaws, a skull fragment, and isolated teeth, there is not enough material to decide whether these fossils belong to the same species as *Homo antecessor*.

Age

No fewer than five different absolute dating methods have been employed to date the various Sierra de Atapuerca cave deposits and the fossils and tools found in them. Paleomagnetic analysis of the Gran Dolina Cave sediments shows a reversal in Earth's magnetic pole during deposition of the TD7 layer. This reversal is known to have occurred 780,000 years ago. Layers TD3-TD6 were all deposited when Earth's magnetic pole was reversed and are thus older than 780,000 years. Found in the top 15 cm of the TD6 layer just under the TD7 layer, *Homo antecessor* fossils are thus slightly older than 780,000 years. This date was confirmed by two other absolute dating methods and by the presence of voles and other small mammal species in the TD6 layer specific to a time interval of 780,000 to 857,000 years ago. Estimated dates in years for the other Grand Dolina layers are as follows: 1 million (TD3-TD4); 900,000 (TD5); 700,000 (TD7) 600,000 (TD8); 390,000 (TD10); and 350,000 to 250,000 years (TD11).

The Ternifine human fossils have no absolute date. They are estimated to have an age of seven hundred thousand years based on animals found in the same deposits.

Tools

There are approximately three hundred tools found in the TD4–TD6 layers of Grand Dolina. Most of these are found in the same deposits as the *Homo antecessor* remains. There are hammerstones, flakes, and cores. All flakes are struck on a single side and some may be retouched. Microscopic analysis of tool wear shows these generalized tools were used for de-fleshing, wood working, and, less commonly, skin preparation. Generally, tools seem to have been made as needed and were not used over and over again or specialized in function. Microscopic wear, however, indicates a relationship between denticulated (finely notched) tools and wood working. Compared to tools found in more recent deposits at Sierra de Atapuerca, a lower proportion of tools associated with *H. antecessor* are made of fine grain material, and higher proportions are made of flowstone. Cut marks on bones made with tools and broken bones with evidence of hammerstone impact suggest *H. antecessor* used tools to de-flesh animals and also extract marrow. The fact that the *H. antecessor* bones found in the TD6 layer have unmistakable marks from stone tools, and are all broken in the same manner as are those of other animals regularly eaten, is suggestive of cannibalism.

Animals and Habitats

Mammals found in the TD6 layer and other Sierra de Atapuerca contemporaneous cave deposits are modern in appearance. Only one, a deer (*Eucladoceros*), belongs to an extinct genus. With the exception of an elephant that is too incomplete to identify, all the other mammals are modern in appearance. These include a hedgehog, a number of shrews, desmans, Old World moles, a few bats, Old World rabbits, hares, porcupines, woodchucks, beavers, a variety of voles (e.g., water vole, harvest vole, and meadow vole), hamsters, garden door mice, Old World field mice, Old World harvest mice, lynxes, laughing hyenas, foxes, wolves, weasels, brown bears, steppe rhinoceroses, horses, pigs, fallow deers, red deers, and bison. Many of these mammals, especially the small mammals, still live in the area today, although they belong to different species from those that existed in TD6 times. Laughing hyenas, horses, and two horned rhinoceroses no longer exist in Europe, and bisons are nearly extinct. Significantly there are no large cats or primates in the TD 6 layer, although the scimitar cat (*Homotherium*) was found in the earlier TD5 layer and existed in Europe as late as 450,000 years ago.

Pollen analysis of the TD 6 layer indicates forests made up largely of junipers, oaks, and live oaks. In the upper part of the TD6 layer associated with *Homo antecessor*, there is a relative

Eight hundred thousand years ago, a dry cave offered safety and shelter to *Homo antecessor* from the harsh and unforgiving elements.

reduction of juniper and an increase in the warmer Mediterranean species such as wild olive and mastic trees (a plant used for making varnish related to the pistachio). There are also beech, hazelnut, and chestnut trees, all of which become more common in the TD 7 and TD8 layers, suggesting a change to more humid conditions. The nuts from these trees and from those of oaks would have been a source of food for *H. antecessor*, as would have been most of the mammals in the deposit. The presence of hamsters, field and harvest mice, heather voles, steppe rhinoceroses, and horses indicates open grassy fields. As occurs today, woods and shrubbery would have existed along ridges and stream courses and intermittently in low areas where water accumulates. Grassy fields would predominate in low areas with little water and on ridges with thin soil. The presence of wild olives and live oaks indicates temperatures rarely dipped below freezing.

Climate

At approximately forty-two degrees latitude north of the Equator and 14 km west of Sierra de Atapuerca (1079 m above sea level), Burgos (890 m above sea level) records mild seasonal temperatures. The coldest and warmest mean monthly temperatures are in December and January (2° C, 35° F), and July and August (18° C, 64° F). Temperatures may dip below freezing from November–March. Mean annual rainfall (572 mm) is spread out throughout the year, with most months receiving at least 46 mm of rain. The exceptions are the two warmest months, July (25 mm) and August (23 mm). Winter precipitation may fall in the form of snow, especially in the mountains. European glacial periods did not directly affect the Iberian Peninsula as they did the rest of Europe. During glacial periods, however, the weather appears to have been wetter and the summers possibly cooler. Regardless, there is no evidence that the climate was drastically different when *Homo antecessor* inhabited the area than it is today.

Classification

If the Ternifine jaw bones are indeed the same species as *Homo antecessor* and this species proves to be distinct from *Homo heidelbergensis*, then the *H. antecessor* species name would be changed to *Homo mauritanicus*. The latter is the species name given to the Ternifine jaws in 1954, and has priority over *H. antecessor*.

Considering that *Homo antecessor* may be only 50,000 years older than the Mauer jaw (the *Homo heidelbergensis* holotype specimen; *see Homo heidelbergensis*), arguments that it is a species distinct from *H. heidelbergensis* must emphasize significant differences between the two. The *H. antecessor* lower jaw fragments (ATD 6-5 and ATD 6-96), however, are from a child and a young adult female. Thus many of the differences claimed to distinguish *H. antecessor* from the fully adult Mauer jaw may be due to sex or age differences. Age differences may also underlie claims, based on the ATD 6-69 child cheek bone, that *H. antecessor* had a more human-like mid-face when compared to Neanderthals and *H. heidelbergensis*.

Although the more modern human-like *Homo antecessor* skeleton when compared to the Neanderthal and the Sima de los Huesos *Homo heidelbergensis* skeleton is impressive, analogous differences may be found in comparisons of Sudanese natives to Inuits (Eskimos). These differences in large part reflect adaptations to climate (*see Homo sapiens*) and may fluctuate during a lineage's evolution. Because such adaptations to climate may appear within a population in a relatively short time, they are not convincing indicators of species differences and/or of ancestor-descendant relationships.

Assuming that *Homo antecessor* is a distinct species, its relationship to some un-named east African skull- and braincases from Olorgesailie, Kenya (KNM-OL 45500), the Dakanylo Member of the Bouri Formation, Ethiopia (BOU-VP-2/66), and the Dandeiro sedimentary basin near the village of Buia, Eritrea (UA-31), may show whether any of these fossils also belong to *H. antecessor*. With smaller brain volumes (750 cc to 995 cc) than *Homo heidelbergensis* and existing within a 900,000- to 1,000,000-year interval, predating *H. heidelbergensis* and *Homo rhodesiensis*, these east African fossils may all prove to be the same species as *H. antecessor*. The same applies to the OH 9 braincase and the OH 12 skull fragments from Olduvai Gorge, Tanzania. Although the OH 9 braincase is believed to come from the Olduvai upper Bed II, dated at approximately one million four hundred thousand years, and the OH 12 skull fragment is believed to come from Bed IV, dated at between 750,000 and 1 million years, both are surface finds that could have come from the same million year time range. Notably the OH 9 braincase shares with the Bodo *Homo rhodesiensis* skull widely separated eye sockets, strong development of the brow ridge, and other skull and face features, suggesting a possible ancestor/descendant relationship. All of this would have important implications as to the ancestor/descendant relationships of *H. antecessor*.

Considering its generalized skeletal features and confirmed antiquity, *Homo antecessor* could be the ancestor of all subsequent European humans (i.e., *Homo heidelbergensis* and *Homo neanderthalensis*). The ancestral status of *H. antecessor*, however, is not supported by newly acquired features shared with later European species to the exclusion of *Homo habilis* or *Homo ergaster*. As such, the likelihood that it is such an ancestor depends on not finding other fossils that share newly acquired features with later European *Homo* species and are contemporaneous with or predate *H. antecessor*. The recent claim based on the ATD 6-96 jaw bone dimensions that *H. antecessor* shows a closer relationship to *Homo pekinensis*, and is the ancestor of modern humans, but not of *H. neanderthalensis*, emphasizes just how little evidence is needed to rethink and overturn past claims.

Historical Notes

In 1962, J. L. Uribari, a member of the Edelweiss Speleological Club, notified the director of the Burgos Museums, B. Osaba, that there were fossil out-cropping's from the railroad cut through Sierra de Atapuerca. In 1968, Miguel Crusafont, director of the Paleontology Institute of Sabadell, Barcelona, sent Narciso Sanchez to collect fossils in the railroad cut, which were taken to the Sabadell Institute. This excavation was followed by those of Trinidad Torres in Gran Dolina and in the Galeria complex in August of 1976. In 1980, the Gran Dolina and Galeria deposits were finally cleaned of surface vegetation, and excavation of the TD11 layer was begun by Emiliano Aguirre and his group. In July of 1994, the first remains of *Homo antecessor*—a lower (ATD 6-5) and upper jaw (ATD 6-13) and eleven isolated teeth—were found by the excavation team that succeeded Aguirre. In a May 1997 issue of *Science*, Jose Maria Bermudez de Castro and colleagues made the jaw fragments (ATD 6-5 and ATD 13) and the isolated teeth (believed to belong to one individual) the holotype specimen of their new species, *Homo antecessor*. The species name "*antecessor*," which in Spanish means "ancestor," is from the Latin word "to come before (explorer)." It was chosen to emphasize that this species is the earliest known European and a likely human ancestor. In 1999, the *Journal of Human Evolution* dedicated a special issue (volume 37) to the Grand Dolina *H. antecessor* finds. In an April 2005 issue of the *Proceedings of the National Academy of Sciences*, Eduald Carbonell and colleagues, reporting on a new found adult lower jaw fragment (ATD 6-96), disputed the original claim that *H. antecessor* could be ancestral to Neanderthals, but stuck to the claim of modern human ancestry. All fossils are housed in the National Natural Sciences Museum in Madrid and in the Archeological Museum of Burgos, Spain.

HOW TO GROW A SUGARPLUM TREE

The harsh rattling chatter of a honeyguide bird high in the crown of a sugarplum tree seemed to beg for someone to follow it. Those willing to do so would no doubt be rewarded with honey. It was late November, and the sugarplum trees were full of fruit. Thus a man returning to his camp, who heard the calls, ignored them, preferring to concentrate on the bountiful fruit supply instead. He unrolled a bark cloth he was carrying, laid it flat on the ground, and filled it with sugarplums. Bringing the ends of the cloth together and tying them off, he picked up the package, placed it on his head, and continued on his way. Maybe tomorrow he would go after honey, but not today. There was no need to follow the honeyguide. He knew of a tree hollow where bees always returned to nest. Every year he would go there to harvest honey.

Some weeks passed before he once again thought about honey. Seeing a honeyguide bird in a tree close by, in the early morning, reminded him. He gathered some things he needed in a pouch he slung over his shoulder, called out to his son, and both set off in search of honey. At noon they crossed a stream and saw eland tracks in the wet sand that flanked the stream. Long ago he had hunted many eland here, but he had not seen their prints for some time. He was surprised to discover that the eland had returned. He followed the eland's trail through some woods and out into the open grassland to see if he could get a glimpse of the animal, but was unable to. He doubled back to look at the trail more closely. The hoof prints were clear of fallen leaves and twigs, and the leaves, on a branch the eland had broken when it walked by, were still green. There was no doubt the trail was fresh. He called out to his son, who was up in a marula tree collecting fruit. They sat under the tree eating the fruit and making plans to return for the eland.

By mid-afternoon they reached the hollow tree trunk. As the man had expected, hundreds of bees were flying in and out of the nest. The man leaned two long and sturdy branches against the tree trunk, and climbed up to get a better look. He smiled. The trunk was full of honey. Climbing down, he found a place to sit. He pulled a small wood block and dowel out of his pouch. He stuck the dowel in a shallow hole in the wood block and rolled it back and forth between his palms. As he did so, his son placed a clump of fine dry grass where the dowel met the wooden block. Gentle blowing on the dry grass soon ignited a flame. The man and his son added more dry grass and some twigs until they had a fire going. The man then smothered part of the fire in a large clump of foliage, picked up the smoldering clump, and stuffed it in the bee's nest. A trail of thick black smoke erupted from the hollow trunk and slowly descended to the ground. After some minutes passed, he placed another smoldering clump of foliage in the tree trunk and went with his son to look for leaves in which to package the honey.

The man laid out the large leaves of a wild banana plant flat on the ground as his son climbed into the smoking bee's nest, which no longer showed signs of life. Emerging with a panel of honey in each hand, he passed the panels down to his father. Cleaning off the clinging bees, the man wrapped each panel in its own banana leaf. In a short time, the nest was cleared, and all the honey packaged, save for a small panel the father and son shared. Carrying the packages full of honey on their heads, they set out for their camp. On the way, they passed a grove of sugar plum trees where a troop of baboons were feeding. Baboon droppings were everywhere. With a stick the man broke apart the droppings and collected the seeds. He wanted to plant some trees close to his camp. He was well aware that the seeds of the sugarplum tree are more likely to germinate once they pass through the gut of a baboon.

HOMO RHODESIENSIS

Skull, Teeth, and Diet

Homo rhodesiensis is known from three skulls, each found at different localities at a considerable distance from each other (*see* map). The Kabwe skull is nearly complete, missing only a portion of the skull base. The Saldanha skull consists of only the braincase, with no face or skull base, and the Bodo skull, with only the face and the top of the braincase.

In many respects, these skulls are similar to those of *Homo heidelbergensis*. As characteristic of *H. heidelbergensis*, the bridge of the nose in both the Kabwe and Bodo skulls is markedly depressed and very wide, widely separating the eyes. In Kabwe, however, the nose opening is smaller, and the cheek bones are narrower, attaching much higher on the upper jaw than in either *H. heidelbergensis* or Bodo. The mean braincase volume for the three individuals is 1267 cc (range 1325 cc to 1225 cc). The skull bone is characteristically thick.

None of these fossils preserve jaw bones, and only the Kabwe remains preserve teeth. Unfortunately, these are too heavily worn with many dental caries for comparing measurements. An upper jaw fragment of another smaller Kabwe individual preserves only a wisdom tooth and a fragment of the second molar, so it too has a limited value for comparisons.

The Kabwe palate is large and, as in most *Homo* species, the teeth set in a parabola. The open skull sutures indicate a relatively young adult. Its heavy tooth wear suggests this individual ate abrasive foods. These may have consisted of rootstocks, tubers, and grains. Otherwise the diet would have been similar to that of *Homo heidelbergensis*.

Skeleton, Gait, and Posture

At Kabwe two-thirds of the lower half of a right arm bone, a right and left hip bone, the lower portion of the backbone articulating with the pelvis (sacrum), a right and left thigh bone, and a complete left shin bone are all preserved. Although missing part of the intervening shaft, the left thigh bone preserves both the knee and hip joint surface, and the right hip bone is almost complete. In most respects , the bones are similar to those of modern humans. The shin bone joint surfaces, however, were oriented in such a way that the foot would have been markedly pigeon-toed when the knee cap faced forward. Overall the long bones appear to have relatively thick shafts, thicker than those of Africans presently living in the same region, but within the modern human range. It is not known how many individuals are represented by the skeletal remains, so that relative proportions of upper to lower limb bones may not correspond to the same individual. Assuming human proportions and based on shin bone length (416 mm), the Kabwe individual would have been 165 cm tall. The lower arm bone fragment found with the Bodo skull has yet to be reported in detail.

Fossil Sites and Possible Range

Homo rhodesiensis is known from a limestone cave in the now defunct Broken Hill lead and zinc mine

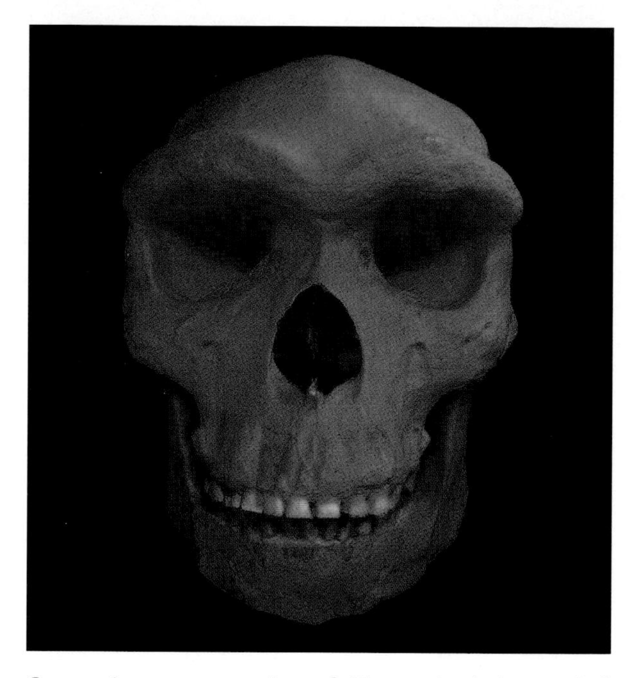

Composite reconstruction of *Homo rhodesiensis* skull based on the Rhodesian Man Broken Hill 1 skull from Kabwe, Zambia, with a modified Mauer (Heidelberg) jaw.

in the town of Kabwe (Broken Hill), Zambia. Miners tunneling in small hills to look for zinc and lead ores in limestone cavities opened up the cave, bringing the fossils to light. At the time it was found, the cave had no visible external opening. The roof over its entrance and part of its original chamber had collapsed long ago (i.e., cave sediments could be found outcropping from the hill surface), but part of the cave chamber within the northeast side of the hill was still intact. The fossils were found in a layer of sandy clay and yellow clay that formed the cave sediment. The cave sediment had a depth of up to 30.5 m and had been impregnated with zinc ore. The fossils are thus highly mineralized.

Homo rhodesiensis is also known from sand dune deposits 24 km east of Saldanha (about 100 km north of Cape Town), South Africa, and from the upper Bodo Sand unit in the vicinity of Bodo in the Middle Awash River Valley of Ethiopia about 13 km north of Maka and 45 km north of Lake Yardi (*see Australopithecus garhi* and *Australopithecus afarensis*).

Age

Neither the Kabwe nor Saldanha fossil deposits have been dated. The Saldanha braincase was a surface find in sand dunes without a known deposit of origin. Remnants of the limestone hill in which the Kabwe remains were found still exist, but no one has yet succeeded in dating the sediments. A volcanic ash layer in a neighboring deposit, supposedly continuous with sediments underlying the Bodo fossils, yielded a radioisotope date of 640,000 years. This date was tempered by archaeological remains and associated animals, and the skull is estimated to be six hundred thousand years old. Guesstimates for Saldanha and Kabwe of between two hundred thousand and seven hundred thousand years encompass the Bodo age.

Tools

Tools associated with the Kabwe remains are made of chert and quartz and are worked on both sides. There is a granite ball that seemed to be shaped into a sphere with worked ivory, bone, and horn. Some of the bone and horn appears to have been used as digging tools. It has been suggested that the sphere shaped ball may have been used for grinding food. The Saldanha dunes and the Bodo deposits both have a rich tool assemblage, which includes bifaces (hand axes) and specialized tools. Cut marks on the Bodo skull suggest it was defleshed after it died. It is not certain whether this was done as part of a burial ritual or for food.

Animals and Habitats

The animals found with *Homo rhodesiensis* are all modern in aspect and most still live in the surrounding area today. A species of serval and of white rhinoceros, very similar to their modern counterparts, are the only animals reported for the deposit that today are extinct. There were shrews, elephant shrews, gerbils, porcupines, and a variety of mice and rats. Predators included leopards, hyenas, jackals, mongooses, and the serval. Hyenas were the most common predators, leading to the

Homo rhodesiensis

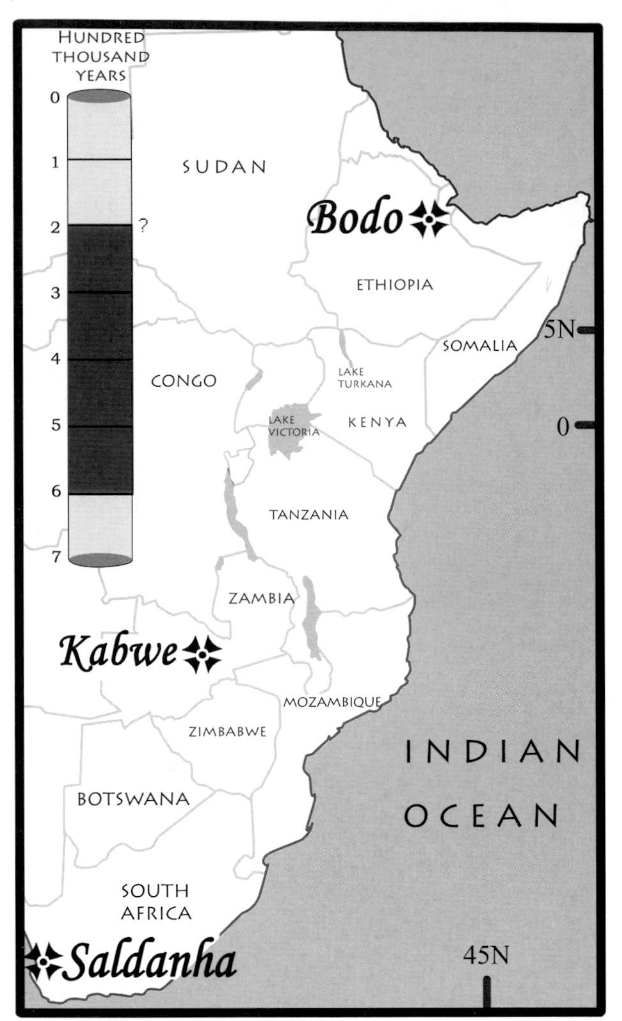

suggestion that the Kabwe cave had been a hyena den. Grazers and browsers include zebras, warthogs, gnus, elands, kudus, Cape buffalos, giraffes, elephants, and two species of white rhinoceros (a modern and an extinct species). Of these, warthogs, gnus, buffalos, and elands were the most common. There were no primates found in the deposits. All animals suggest a habitat that is very much like the one that exists there today: open miombo woodland interspersed with grass-covered wetlands.

With a much richer fauna, the Saldanha deposits record ratios of living to extinct genera of approximately five to one. Extinct genera include the dirk-tooth cat (*Megantereon*), the short-necked giraffe (*Sivatherium*), an extinct bush pig (*Kolpochoerus*) and warthog (*Metridiochoerus*), an extinct buffalo (*Pelorovis*), two extinct hartebeests (*Parmularius* and *Rabaticerus*), and the giant hartebeest (*Megalotragus*). Most species still live in the surrounding area. Considering the nature of the deposit (i.e., shifting sands), it is likely that it may also be recording animals that were not contemporaneous with *Homo rhodesiensis*.

Climate

At 1207 m above sea level and fourteen degrees south of the Equator, Kabwe records temperatures that vary very little during the year. Mean monthly temperatures are lowest for June (15.6° C, 60° F) and highest for October (24° C, 75° F). At 8° C (46° F) June also records the lowest mean monthly minimum. Mean annual rainfall is 907 mm, with nearly 75 percent of the total annual rain falling in December through February. Over an eighty-nine-year period, June, July, and August received no rainfall, and May and September means combined for only 5 mm of rainfall.

Saldanha receives 508 mm of rain annually, more than 60 percent (318 mm) falling in the winter months of May–August. November through February receive only 10 percent (51 mm) of the yearly rainfall. Mean monthly temperatures are lowest for July (45° F) and highest for January (70° C). Rarely do the temperatures dip below freezing. There is no evidence that rainfall and temperatures were much different when *Homo rhodesiensis* lived in these areas than they are today. (*See Australopithecus garhi,* Climate for Bodo.)

Due to an increasing wanderlust, *Homo rhodesiensis* sometimes found themselves in unfavorable conditions. This male has been lost for days on the desert, with no water or shade in sight.

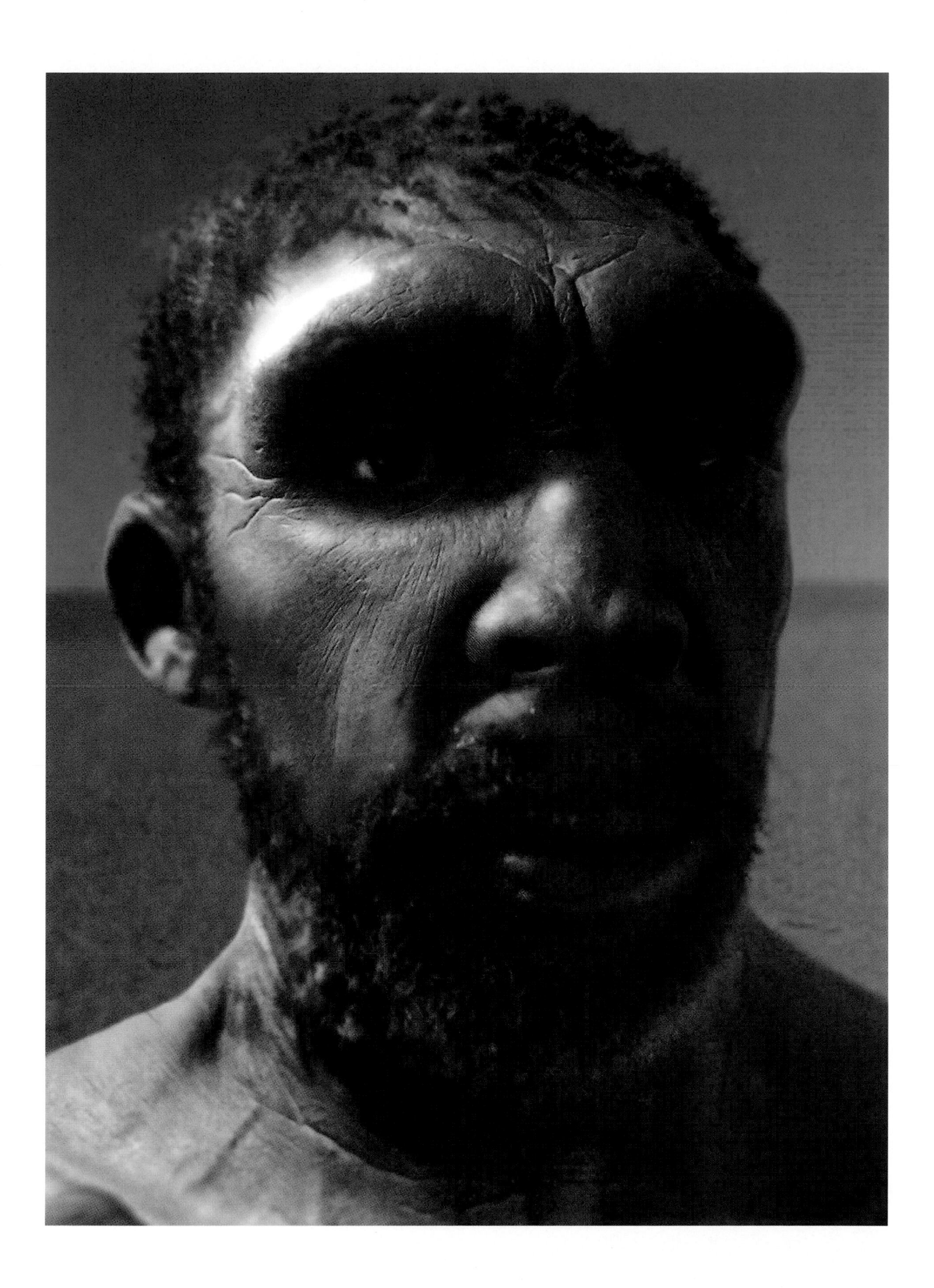

Classification

Based on what exists of the skull, it is almost certain that *Homo rhodesiensis* and *Homo heidelbergensis* are members of the same species. Detailed comparisons of Kabwe to Sima de los Huesos skeletal bones, however, are necessary before making a decision. The most characteristic *H. heidelbergensis* and *H. rhodesiensis* fossils are male. If the small upper jaw bone found at Kabwe is any indication, females may have been very small. Females, therefore, may not be as easy to identify, especially given the possibility of more than one *Homo* species occurring in Africa at this time. It still remains to be seen if males and females differ as much in size in *H. rhodesiensis* as in *H. heidelbergensis*.

Historical Notes

An anonymous Zambian miner is credited with finding the Kabwe skull in 1921. In a November *Nature* issue that same year, Arthur Smith Woodward reported on the skull, making it the holotype specimen of *Homo rhodesiensis*. In a 1928 *British Museum of Natural History* publication, William Pycraft and colleagues described the fossils and archaeological remains found in the Kabwe Cave.

Keith Jolly and Ronald Singer are credited with discovering the Saldanha braincase in 1953. Singer reported on this find in a 1954 issue of the *American Journal of Physical Anthropology*. In a special 1958 volume on Neanderthals edited by Gustav von Koenigswald (*Hundert Jahre Neanderthaler*, Kemink en Zoon), Singer pointed to differences between the Saldanha skull and those of European Neanderthals.

Paul Whitehead and Alemayehu Asfaw are credited with finding the Bodo skull in November of 1976. At the time that it was found both *Nature* and *Science* refused to publish a report of its discovery. Additional skull parts were later found by Desmond Clark in 1981. In 1992, the Bodo skull's description was the subject of a New York University Ph.D. dissertation by Tsirha Adefris. This was followed by Philip Rightmire's comparative description in the *Journal of Human Evolution* in 1996.

The Kabwe, Bodo, and Saldanha remains are housed in the British Museum of Natural History, London, the National Museum of Ethiopia, Addis Ababa, and the South African Museum, Capetown. The species name *"rhodesiensis"* refers to the British colony of Rhodesia, named after Cecil B. Rhodes, the owner of the British East India Company.

GREED AND EQUALITY

He hummed quietly to himself as he searched among the stones he had collected over the years. He wanted one that was just right to make a new blade for his spear. He reached for a big block of jasper and angled it so that one of its small sides faced up. Striking the edge of the face with a large hammer-stone, he produced a large, long flake. He picked up the flake, brought it out into the sunlight and turned it over and over in his hands. He repeatedly hit the flake on both sides with a small stone, until it finally acquired the shape he sought. Satisfied with his handiwork, he sat until sunset examining and retouching his newly made blade.

That night he worked by the fire using the skin and guts of a boar and the resin from a mastic tree to haft the blade to a wooden shaft. When the blade was finally hafted, he rolled the spear around in his right hand, pretending to throw it, and finding the point along the shaft at which it balanced.

The next morning his uncle, a brother, and a cousin woke him up before sunrise, and all four left camp for the river valley. His brother had recently seen a large herd of bison and knew where they were pasturing. By noon they met up with the herd. His uncle and his brother would try to run the animals between some large boulders flanking a dry river bed. He and his cousin would wait behind the boulders and spear a bison as the herd ran by. After repeated efforts, the herd started moving in the direction of the river bed. A large cow and some bulls passed first, but wanting to avoid turning back the herd, he and his cousin waited. They both threw their spears into an animal midway in the herd. In his eagerness to do so, his cousin had stepped too far in front of the boulder and his leg was trampled by the herd. He was alive, but could not walk.

His uncle stayed with his cousin, while he and his brother went in search of the wounded animal. The animal had separated itself from the herd, taking refuge in a small patch of shrubs. As he distracted the animal, his brother climbed a tree. Shimmying out along a branch, his brother took a position above the bison and speared it from above between its shoulders. The bison lost its strength and dropped in convulsions. He wrenched his brother's spear out and thrust it back into the bison, until it stopped breathing. His uncle, carrying his cousin, soon arrived and helped butcher the animal. Since the three of them could not carry his cousin and the bison, his uncle left to get help.

An assortment of children arrived to help. Behind them came four men carrying spears. The four had heard of the kill and demanded a percentage for the owner of the herd. Their uncle argued that no one owned the herd or the land, but with their cousin unable to fight, they grudgingly handed over nearly half.

As he retuned to camp, taking turns to carry his cousin on his shoulders, he became increasingly angry at loosing half of the bison. It hadn't been the first time it happened. His wife's father was always claiming ownership and taking at least half of what he worked hard to get. He had been warned many times that this man would always take from him, but when he and his wife had first come together he was young and all that mattered in the world to him was her. Seeing his cousin pale and in pain, made him even angrier. He thought about how happy he and his family would be if those who took advantage of the sacrifice of others, claiming ownership of things that no one owns, would all be dead.

The anger of having his half the bison stolen made him loose his appetite. He did not eat anything for the remainder of the day and at night he could not sleep. Getting up after the moon descended, he walked to his father-in-law's camp. The old man lay sleeping under the stars, uphill from his two youngest wives who lay on a bear skin below a sprawling live oak. He thrust the blade of his spear still stained with the blood of the bison through the center of the old man's chest. The old man's snoring gave way to a muffled gurgle and blood pooled in his chest and dripped out his mouth. He silently carried the old man uphill to an open shaft. He severed the head from the body, tying it by the hairs to a sapling. He dropped the body down the shaft. It was the same shaft the old man had thrown dozens of dead youngsters including a stepson and a grandson, after making their parents witness their death. Claiming the youngsters were incorrigible thieves who repeatedly stole from him, the old man had convinced everyone that killing them was just punishment and assure parents would control their other children.

By the time he reached his camp, he was exhausted, but was at peace with himself. He lay down next to his wife, put his arms around her swollen stomach, and remembered why he loved her so much.

HOMO HEIDELBERGENSIS

Skull, Teeth, and Diet

Most *Homo heidelbergensis* remains are very incomplete and consist of partial skulls or skull parts, including the holotype specimen lower jaw from Mauer (near Heidelberg, Germany) on which the species name is based. Most of what is known about *H. heidelbergensis* comes from Sima de los Huesos, 14 km east of Burgos, Spain (*see Homo antecessor*), where remains of approximately thirty-two individuals have been uncovered.

As seen in the most complete skull (AT-5), there is a well-pronounced single bone bar ("unibrow") above the eye sockets, displaced downwards at its midline. The eye sockets are widely separated by a very wide nose bridge. The nose bridge is depressed inwards relative to the bulging brow above and the protruding nose bones below. The lower edge of the eye sockets is set slightly above the upper border of a very large nose opening. The cheek bones attach on the upper jaw at a level between the first and second molars. They are relatively wider from top to bottom than in Neanderthals and thus attach lower on the upper jaw. Relative to the cheek bones, the nose and upper jaw jut out considerably, forming a face with somewhat of a snout. As in Neanderthals, the braincase is elongated and may exhibit a "bun" shape at the back. The forehead, however, is generally lower than in Neanderthals, and the braincase behind the eye sockets narrows more than in Neanderthals. Mean braincase volume for ten individuals is 1274 cc (range of 1116 cc to 1450 cc), slightly less than in Neanderthals and modern humans. Skull base and ear canal measurements suggest *Homo heidelbergensis* was sensitive to the same sound frequencies as modern humans.

Homo heidelbergensis upper and lower jaws are large. The lower jaw may have a space behind the last molar as in Neanderthals. On average cheek teeth are wider from side to side and the lower front teeth broader from front to back than in Neanderthals. The teeth, especially the canine, and the lower jaw appear to show greater size differences between males and females than they do in modern humans. Compared to modern humans, there appears to be early development of the wisdom tooth and a delay in canine development.

Microscopic studies on *Homo heidelbergensis* tooth wear at both Arago, France, and Sima de los Huesos, Spain, suggest an abrasive diet composed of at least 80 percent plant material, as is common in modern human hunter and gatherer societies. Wider cheek teeth than Neanderthals, which indicate more roughage in the diet, also support this interpretation. As in modern humans and Neanderthals, tools associated with *H. heidelbergensis* suggest some degree of food preparation. Unlike in Neanderthals and *Homo antecessor*, there is as of yet no strong evidence of cannibalism.

Skeleton, Gait, and Posture

The *Homo heidelbergensis* skeleton is very similar to that of Neanderthals. The pelvis from Sima de los Huesos is nearly indistinguishable from that of Neanderthals. Because no associated remains of *H. heidelbergensis* are known, the Sima de los Huesos material is badly broken up, and the Ehringsdorf material is too incomplete, it is not clear how its body proportions differ from Neanderthals. Long bone lengths and mid-shaft cross-sections suggest limb proportions were very Neanderthal-like. What is known of the hands and feet also show proportional similarities to Neanderthals.

Assuming modern human proportions and based on a complete arm bone (AT-787, AT-788, and AT-1115) from Sima de los Huesos, *Homo heidelbergensis'* stature is estimated at 173.1 cm. Making the same assumptions, a 175.3 cm stature is estimated for the Boxgrove shin bone. Given a lower skull than modern humans, however, their height would have been 2.5 cm to 3.7 cm shorter. Although their teeth show greater size differences than modern human male and females do, skeletal differences between the sexes are comparable to those seen in modern humans.

Fossil Sites and Possible Range

Homo heidelbergensis is known mostly from limestone caves and mines and river gravel beds in Europe at elevations below 1000 m in the following countries: England, France, Spain, Germany, Greece, Hungary, and Italy. It is also known from Israel and Morocco. The northern limit of its range is in Germany at approximately fifty-one degrees North latitude. Zuttiyeh Cave in Israel on the northwestern shore of Lake Kinneret (the Sea of Galilee) marks its southern and eastern limits. The limit of its western range is at Thomas Quarry, Morocco. Although it is more conservative to assign *Homo rhodesiensis* from Kabwe, Zambia, to a separate species, given the intervening Sahara Desert and forty-seven degrees of latitude (5300 km) between Kabwe and the southern limit of *H. heidelbergensis,* the two are likely to be the same species. The Narmada braincase from India and the Dali skull from China may also belong to *H. heidelbergensis.*

Age

Classically the *Homo heidelbergensis* deposits have been dated by correlation to European glacial periods. They are believed to predate Neanderthal deposits and are found within the first half of the Riss glacial and the second to last (penultimate) interglacial (i.e., the Holstein or Mindel-Riss). Still to this day, most remains lack absolute dates. Most dates are thus guesstimates based on correlation to glacials or other events, and on anatomical similarities to classic European Neanderthals. The more similar they are to Neanderthals, the more recently they are believed to have occurred. *H. heidelbergensis* sites with such estimated dates in years are as follows: in England, Swanscombe 250,000 to 350,000 years, and Boxgrove 500,000 years; in France, Arago 450,000 years, Lazaret 200,000 years, and Mountmaurin 200,000 to 300,000 years; in Germany, Mauer 500,000 to 735,000 years, Reilingen 125,000 to 250,000 years, and Steinheim 225,000 years; in Israel, Zuttiyeh 250,000 years; in Italy, Casal de' Pazzi 200,000 to 250,000 years, and Ceprano 460,000 to 880,000 years; and in Morocco, Thomas Quarry 400,000.

Homo heidelbergensis absolute dates are known for fossils at the following sites: in France, Bau de l'Aubesier 190,000 years; in Germany, Bilzingsleben 300,000 to 400,000 years, and Ehringsdorf 200,000 to 250,000 years; in Greece, Petralona 150,000 to 250,000 years; in Hungary, Vertesszollos 250,000 to 300,000 years; and in Spain, Sima de los Huesos 350,000 years. Various methods relying on sediment, bone and teeth (some time from the fossil itself), and/or associated stone tools have been used to arrive at these absolute dates.

Tools

Homo heidelbergensis tools are manufactured using a prepared core technique (e.g., Levallois), as are those of Neanderthals. Usually the rock used is very fine grained, and tools may be small with fine flaking. Flakes struck from stone may be worked on both sides. Tool kits are varied. They include those tools associated directly with butchering and consuming animal meat and also tools used for wood working and skin preparation. Specialized tools

Homo heidelbergensis

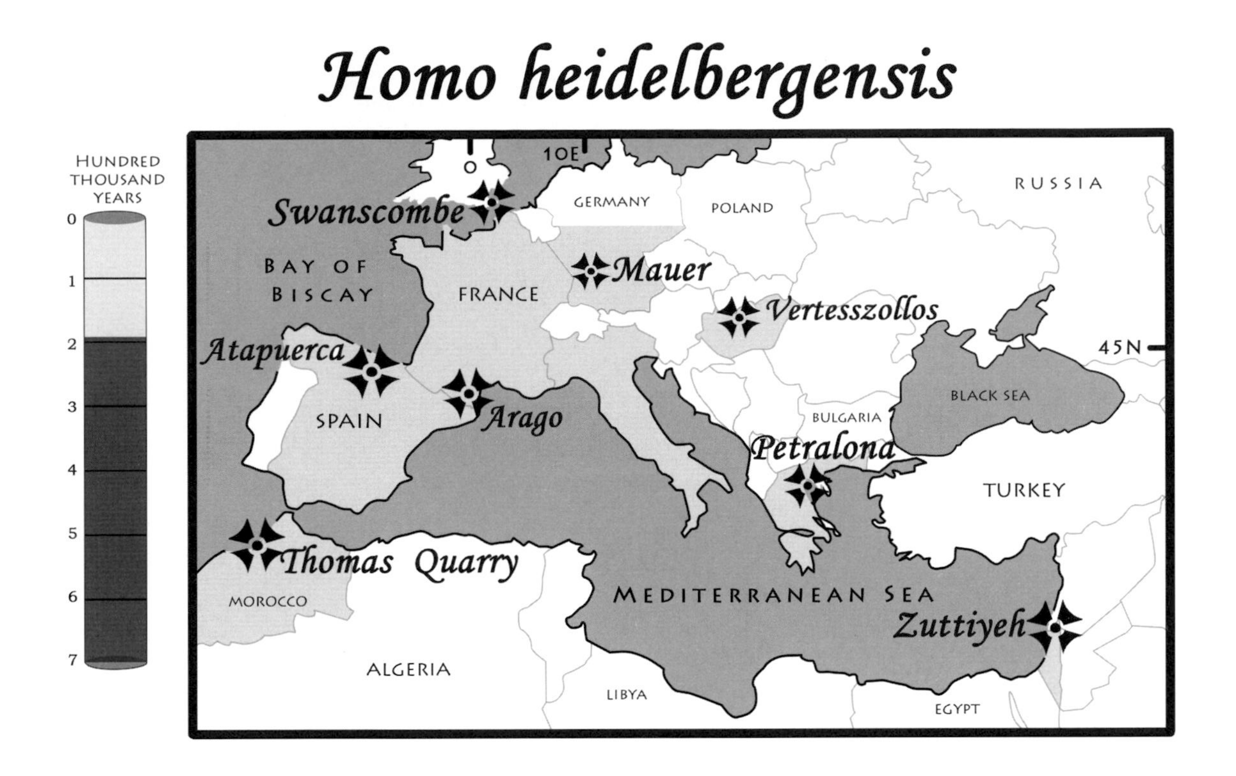

appear to have long term use and are not immediately discarded after being used once or twice. In contrast to modern humans, decorative objects have not yet been reported for *H. heidelbergensis*.

Scratches on upper and lower front tooth enamel, seen to coincide when jaws are closed, suggest *Homo heidelbergensis* at Sima de los Huesos was using stone tools to cut material it held between the teeth. Because most scratches progress from the upper left to the lower right of the tooth face, most *H. heidelbergensis* individuals at Sima de los Huesos were probably right-handed. Some *H. heidelbergensis* front teeth exhibit beveled wear, indicating the teeth may have been used for preparing skins or in other functions besides eating. In some cases, teeth may exhibit wear associated with toothpick use.

At Sima de los Huesos the presence of only humans fossils in a shaft within a large cave that otherwise contained only carnivore fossils strongly suggests that *Homo heidelbergensis* transported these remains and dropped them down the shaft. A high incidence of females between fifteen and nineteen years old (seven out of thirty-two bodies) and of individuals of indeterminate sex between ten and nineteen years old (twenty out of thirty-two bodies) suggests some type of human selection.

Animals and Habitats

Despite the fact that Sima de los Huesos is one of the richest *Homo heidelbergensis* deposits, the only non-human fossil mammals found in the cave are all carnivores. Most of these are species that still exist today in Europe: the wolf, weasel, marten, European wildcat, and lynx. Others are no longer found in Europe: the lion and the hyena. The only extinct species is a prehistoric relative of brown bears, which is also a relative of the cave bears of Neanderthal times.

Other deposits show a more balanced sample of the mammals that lived with *H. heidelbergensis*. Ehringsdorf, a lime mine near Weimar, Germany, has yielded a relatively rich fauna associated with *Homo heidelbergensis*. The majority of mammals are modern European species not much different from those found in the area today. There are shrews, a beaver, ground squirrels, numerous voles, field mice, birch mice, a European hamster, foxes, wolves, brown bears, European badgers, martens, lynx, wild pigs, fallow deer, red deer, and goats. One species, the laughing hyena, no longer occurs in Europe, but is found in Africa. Extinct species are represented by bison (*B. priscus*), clawless otter (*A. antiqua*), forest elephant (*E. antiquus*), cave bear (*U. spelaeus*), horse (*E. taubachensis*), steppe rhinoceros (*D. kirchbergensis*), and moose (*A. latifrons*). Other modern species belonging to the above genera, however, still occur today either in Europe (*Ursus, Equus, Alces,* and *Bison*), Africa (*Aonyx*), or Asia (*Elephas* and *Dicerorrhinus*). *Megaloceros*, the Irish elk, is the only extinct mammal genus found in the same layers with *H. heidelbergensis*. Diagnostic of the second to last interglacial period, the Ehringsdorf *H. heidelbergensis* fossil deposits are sandwiched between deposits yielding woolly rhinoceroses (*Coelodonta*), woolly mammoths (*Mammuthus*), and reindeers. Animals found at other *H. heidelbergensis* sites, not seen at Ehringsdorf or Sima de los Huesos, include an extinct beaver (*Trogonotherium*), roe deers, chamois, and tahrs.

Trees at Ehringsdorf are also characteristic of temperate areas. There are linden trees (basswood) and a variety of oaks and elms. Secondary growth and shrubs in flooded land or along lakeshores and streams include speckled alder, gray willow, European privet, Hungarian lilac, blood twig dogwood, and hazel nut trees. Many of these trees cannot survive in areas where temperatures dip below -29° C (-20° F), confirming that the temperatures then were not much colder than they are today. The presence of live oaks and wild olives in the Sierra de Atapuerca deposits, likewise, suggests temperatures and a habitat that did not change much over the years.

Climate

Considering it inhabited more or less the same geographic area in Europe as Neanderthals, but at earlier times, the climates *Homo heidelbergensis* endured were probably similar to those endured by Neanderthals. Although dating is not yet accurate enough to be certain, *H. heidelbergensis* must have also lived through glacial periods. At its southern and eastern extent, in Zuttiyeh, Israel, the climate corresponds to that described for Haifa (*see Homo neanderthalensis,* Climate). At Rabat, Morocco, the limit of its western extent, the dry season (June –September) receives less rainfall (18 mm) than in Burgos, Spain (*see Homo antecessor,* Climate), but total annual rainfall amounts are comparable (581 mm vs. 572 mm). Temperatures here would be somewhere between those at Burgos and Haifa. In Weimar, Germany, approximately at the limit of its northern distribution, yearly rainfall is 762 mm distributed more or less evenly throughout the year. February records mean temperatures below freezing, but all the other mean monthly temperatures are above freezing, with the warmest month, August, averaging 18.3° C (65° F).

A high frequency of pine pollen relative to other pollens at Sima de los Huesos during *Homo heidelbergensis* times suggests that the climate may have been drier and possibly colder than it is today. With cold climatic conditions over a large part of its range, *H. heidelbergensis* would have probably used fire and shelter and, in certain areas, clothing.

Classification

Differences between *Homo heidelbergensis* and Neanderthals are best interpreted as those exhibited in evolving lineages over time. In fact, some fragmentary remains (e.g., Biache Saint-Vaast, France) can be comfortably placed within either species. Others, usually the most recent *H. heidelbergensis* fossils, e.g., Swanscombe, Ehringsdorf, Reilingen, Steinheim, Montmaurin, and Sima de los Huesos, all exhibit a number of Neanderthal-like traits. Whether all known *H. heidelbergensis* populations evolved into Neanderthals, or only some that have yet to be sampled did, is an open question.

After a successful yet brutal hunt, a male *Homo heidelbergensis* proudly goes home to the caves of Atapuerca to boast of his ordeal.

In England *Homo heidelbergensis* populations probably became extinct during glacial periods. With loss of a land connection between northern Africa and Spain, *H. heidelbergensis* in northern Africa and Europe would have diverged along different lines. A group based largely on incomplete skull remains, however, is likely to hold some surprises. Although given what is known, all *H. heidelbergensis* fossils may comfortably fit within one species; further fossils may show that more than one species exists or that some are best referred to Neanderthals.

Historical Notes

D. Hartmann is credited with finding the Mauer lower jaw in October 1907 in a rock and sand quarry 1 km north of Mauer, Germany. The following year in 1908, Otto Schoetensack published his description (Engelmann, Leipzig), making the Mauer lower jaw the holotype specimen of *Homo heidelbergensis*. That same year fragments of a human skull were found at Ehringsdorf Quarry. Gustav Schwalbe, in a 1914 issue of *Anatomischer Anzeiger*, based the first report of the Ehringsdorf human finds on a lower jaw found the previous

year. The 1908 finds went undescribed until Gunter Behm-Blancke's 1960 publication on Ehringsdorf.

In September 1959, Christos Sariyanidis found the Petralona skull in a cave just outside of the village of Petralona, Greece. This skull find was reported the following year in a 1960 issue of *L'Anthropologie* by P. Kokkoros and A. Kanellis, claiming it was accompanied by other skeletal remains. The latter have never been found. In 1976, Trinidad Torres, a graduate student, found a lower jaw bone at Sima de los Huesos, which he brought to the attention of Emiliano Aguirre. Aguirre returned to the fossil site and uncovered additional human teeth, jaw, and braincase fragments, and long bones. Aguirre and Torres together with Jose Maria Basabe reported on the finds in a 1976 issue of *Zephryus*, showing they were allied to the Mauer, Montmaurin, and Arago humans. Systematic excavation of the Sima de los Huesos deposit was carried out between 1984 and 1998. In 1997, the *Journal of Human Evolution* dedicated an entire issue (volume 33) to the Sima de los Huesos remains. Fossils are housed in various museums around Europe and in Morocco. The species name "*heidelbergensis*" honors the town of Heidelberg (the oldest University town in Germany) near to where the Mauer lower jaw was found.

This *Homo heidelbergensis* man worries about his cousin who was injured in a bison hunt.

HUNTERS AND HUNTED

He was old now. He no longer had the enthusiasm to go out on the hunt. Hunting was for the younger men, his sons, his grandsons, even some of his daughters, but it was no longer something he wanted or needed to do. His family would care for him. The land was rich with game. He still liked to set snares out for deer and rabbit and, when his sons were not around, to check the pitfall traps they had dug. Occasionally a boar or a small goat would fall in. If it wasn't dead from the cold of the night, he would spear it until it died. If the traps were empty, he was content to sit at the top of the ridge and look down at the game below.

Spotting the animals made him happy, and it made him remember his past. He longed to see those animals he knew as a boy in the faraway land in which he was born. The large herds of antelope and gazelles didn't exist here, nor did the little dassies that cluttered rocky ledges with their pellet-like droppings. He remembered the pungent taste of their meat. He proudly held one in his hand that morning when he heard his sister's screams. He saw the men with the large foreheads and small faces repeatedly hitting her over the head with a stone until she was dead. From where he stood he could see the "little faces" butchering his father and uncle's bodies, cutting their lifeless skin into straps to tie down the meat so they could carry it on their backs. He could see that his little brother was alive. One "little-faced" woman held him in her arms. Although he wondered when they would eat him, he didn't try to save him. He was too frightened, and just turned and ran away in the opposite direction. The "little faces" were too preoccupied with their task to notice he was watching. They were very fast runners, and had they seen him, they certainly would have caught him also. He kept moving until nightfall and found a rock ledge to sleep under. He heard the cries of the dassies twice that night, once before falling asleep and once before morning broke. He wondered if anyone in his family was alive. He wondered if it had happened because he had killed the dassie.

Weeks, maybe months, later, he met up with another boy, and they hunted together, following the herds out into the desert when the rains came. They lived like lions, eating from the herds and never sleeping in the same place. Over the years, they became men, reaching a land green with trees and framed by white mountains. Here there were large herds of deer and bison, and cold nights. Although his skin was warm from the fire and his belly full of meat, he longed to have someone next to him.

He had never mixed with any of the other groups they had met before. Having seen his family killed and butchered made him distrust even those on whom he could depend. This group with which they now shared the fire, however, was different. It was full of young women; their skin was soft and white, not dark like his, and their hair long and silky. He never slept as comfortably as he did that night. With the warmth of their bodies next to him there was no need to be close to the fire.

One day he and his friend parted from the group and went their separate ways. Two of the sisters went with him, and one with his friend. One of his two wives had died before ever giving birth. The other still lived, but was no longer producing children. Over the years he had many more wives, but he never reminisced over them as much as he did over those two.

His last wife was different. He had found her cold and hungry in the mountains, feeding on acorns she was gathering from the ground. Her arm was mangled, probably from an encounter with a boar. When he first met her he had the urge to kill her, but she looked so young, frail, and sickly, he could not bring himself to do it. She smelled different from the women he had been with before, her body was long and slender, and her voice was higher than a child's; she had a circle of scars burned into the small of her back. She had almost died giving him a child, which took hours to be born. The child seemed weak and slow to grow, but was full of energy. He was barely waist-high, and already he was an expert at setting traps. He wondered whether his young son would always hunt animals or one day would also hunt humans.

HOMO NEANDERTHALENSIS

Skull, Teeth, and Diet

The Neanderthal skull is very well known. Many sites have yielded complete or nearly complete skulls: Amud (Israel), Tabun (Israel), La Chapelle aux Saints (France), Mt. Circeo (Italy), Shanidar (Iraq), La Ferrassie (France), Teshik-Tash (Uzbekistan), and Engis (Belgium). Overall, Neanderthal skulls are large with a large braincase. Twenty-four skulls yield a mean braincase volume of 1420 cc with a range between 1200 cc and 1740 cc. This mean value is more or less similar to that of modern humans. The Neanderthal braincase, however, is elongated from front to back and much lower than that of humans with a low forehead. Associated with its elongation, the skull case also juts out in the back, creating a characteristic bun shape.

The face is large and long and not as vertically set as those of modern humans. Eye sockets have a shape reminiscent of aviator glass lenses. Each socket is associated with its own separate but strongly bulging brow ridge. The cheek bones are slender, but the sinuses below them are very large, giving the face a puffy appearance. The opening for the nose is also large and broad. The enlarged sinuses on either side, therefore, jut into the nose opening, just behind its rim.

The lower jaw shows a gap behind the last molar and, in contrast to modern humans, lacks a jutting chin. Both upper and lower jaws are larger than modern human jaws, as are the front teeth. The molars, however, are relatively narrower than modern human molars. Both upper and lower molars have a tendency to develop an extra central cusp not commonly seen in modern humans. Neanderthal front teeth appear to grow at a faster rate and take less time to grow to their final size than those of Upper Paleolithic *Homo sapiens*. Their growth rate, however, appears to be within the range of modern human populations.

The best evidence as to Neanderthal diet comes from analysis of scratch marks and tooth wear. Neanderthals from La Quina (France) and from the Middle East sites of Tabun and Amud (Israel) have tooth-wear similar to Inuits (Eskimos) and other groups that eat mainly meat. Other Neanderthals, St. Césaire (a late surviving population), Malarnaud, and Marillac (all in France) have tooth wear typical of hunter and gatherer groups that are still alive today. This wear suggests a mixed diet of meat with a large helping of fruits and vegetables. Because scratch marks on teeth are erased over time by new wear, differences in wear could reflect seasonal differences in foods eaten and thus time of year the studied individual died. Nitrogen isotope analysis of Neanderthal bone, from three European sites (i.e., Marillac in France, Vindija in Croatia, and Scladina in Belgium), indicates most consumed protein came from land living animals.

Because Neanderthals are assumed to be mainly meat-eaters, given that plant foods are available for only a short portion of the year in cold habitats, it is significant that Middle East Neanderthals living in the relatively warmest climates consume mostly meat. Such a diet may suggest that meat-eating was more than just a preference due to plant food scarcity in winter.

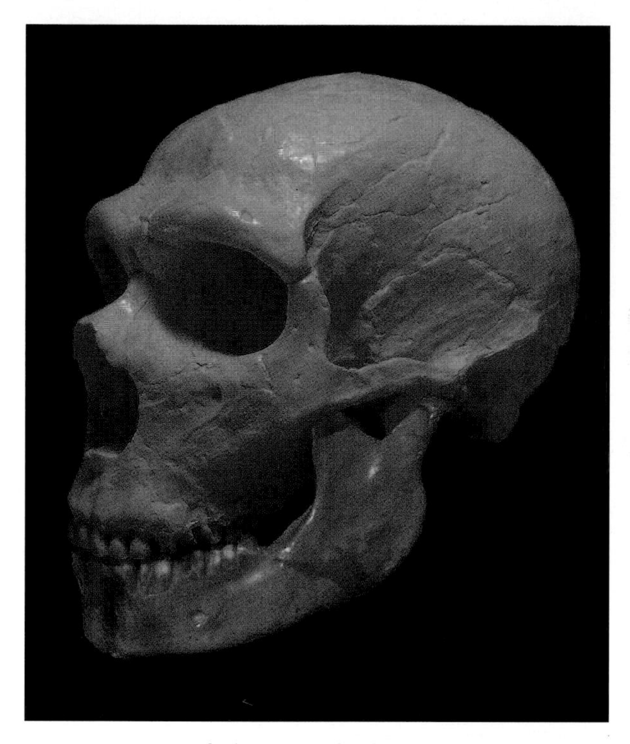

Reconstruction of the Neanderthal La Ferrassie skull, using additional parts from the La Chapelle-aux-Saints, Shanidar 5, Spy 1, Gibralter II, Saccopastore 1, and Saint Césaire skulls.

Archaeological sites show evidence that Neanderthals systematically butchered large game animals and broke bone open to eat marrow. Neanderthals were probably hunters, who may have also herded, managed, but not domesticated their prey. At a large number of sites (e.g., Combe Grenal, Krapina, Macassargues, Marillac, Moula-Guercy, Vindija, and Zaffaraya) evidence suggests Neanderthals themselves were being butchered and eaten. Neanderthal bones from Moula-Guercy, a 100,000-year-old site in Ardeche, southeastern France, show the same breaks and stone tool cut marks as do butchered deer from the same site. This provides the strongest evidence so far that Neanderthals were cannibals. It is unknown, however, whether Neanderthals may have actively hunted each other, as present-day tribes in New Guinea are known to do, or just opportunistically scavenged dead remains of their peers.

Skeleton, Gait, and Posture

Neanderthal backbones, pelvis, shoulder blades, limb bones, and hand and foot bones are all well known. A number of complete or nearly complete skeletons have been excavated (e.g., La Chapelle-aux-Saints, France, and Kebara, Israel) and at most sites other remains, aside from skulls, are commonly present.

Neanderthal skeletons are more or less like those of modern humans. They differ from modern humans mostly in their proportions. Neanderthals have much wider, sturdier built pelvises and relatively stockier lower limb bones than modern humans. By contrast their upper limbs are relatively light in build. Length ratios of forearm to arm and of lower leg to thigh are commonly lower than in modern humans, reflecting the relatively short Neanderthal forearm and lower leg. Their fingers are short and stout. The chest is big, deep, and cone-shaped, and the space between the pelvis and thorax is short. This gives their trunk a stocky and compact appearance, and is in large part responsible for their overall shorter height when compared to modern humans.

Considering that Neanderthals lived during a glacial period, these skeletal differences have been explained as adaptations to cold climates. Neanderthals appear to show more extreme skeletal adaptations to cold than do Inuits. This fact may be best explained by a much longer evolutionary period in cold climates than undergone by Inuits. Other factors, however, that may be responsible for Neanderthal skeletal differences, i.e., heritage, diet, and mode of movement, have not been given much attention. It is possible that one or more of these variables had important causative roles for their skeletal differences. Neanderthals are slightly shorter in height than the modern European population, with more or less the same body weight.

Fossil Sites and Possible Range

Homo neanderthalensis is known mostly from caves, rock shelters, and fissure fillings in Europe and Asia at elevations below 1000 m in the following

countries: Belgium, Croatia, Czech Republic, France, Georgia, Germany, Hungary, Israel, Iraq, Italy, Portugal, Spain, and Uzbekistan. Kebara, a cave deposit on Mt Carmel's southwestern end in what is today Israel, marks the limits of its southern distribution. Feldhoffer Grotto in the Neander Valley, Germany, outside of the Wuppertal suburbs, just east of the Rhine, marks the limit of its northern distribution. Its northern limit decreases eastwards so that it is found approximately below fifty degrees North in the Czech Republic and forty-two degrees North in Georgia. Neanderthal eastern limits are at Teshik-Tash, a cave in southeastern Uzbekistan at a little over sixty-seven degrees North longitude. At 3150 m above sea level, Teshik-Tash also records the highest elevations at which Neanderthals have ever been found. This elevation is considerably above that of the other Neanderthal sites, all of which are less than 1000 m above sea level. Unfortunately, Teshik-Tash yielded an eight-year-old child. Although it is characteristically Neanderthal-like, the fact that Teshik-Tash records both altitudinal and eastern limits raises the question of whether Teshik-Tash adults would also have been as characteristically Neanderthal-like.

If future finds show Teshik-Tash is distinct from Neanderthals, Shanindar east of Mosul, Iraq, and Sakajia west of Tbilisi, Georgia, place Neanderthal eastern limits just past forty-four degrees East longitude. The western limits of Neanderthals appear to be defined by the contours of the European continent and Atlantic Ocean. If, as seems likely, additional finds confirm the molar and premolar found at Columbeira and Figuera Brava, respectively, are Neanderthals, their western limit was in present-day Portugal. Although in the past the Iberian Peninsula was connected to Africa, and the Neanderthal Middle East localities of Amud, Kebara, and Tabun are continuous by land into Africa, there are as of yet no known African Neanderthals.

Age

Classically, European Pleistocene fossils were dated relative to the animal assemblages (i.e., faunas) that coexisted at any one time, and changes in this coexistence (i.e., extinction of one or another species) associated with cyclical expansion and retraction of European glaciers. Neanderthals were no exception. They were believed to appear at the end of the Eemian interglacial and exist throughout the Wurm glaciation, disappearing at its end. The realization that major glaciation events were broken up by one or more warm periods, however,

The play of light and shadow between tree, sun, and sky fills this Neanderthal man with a sense of awe.

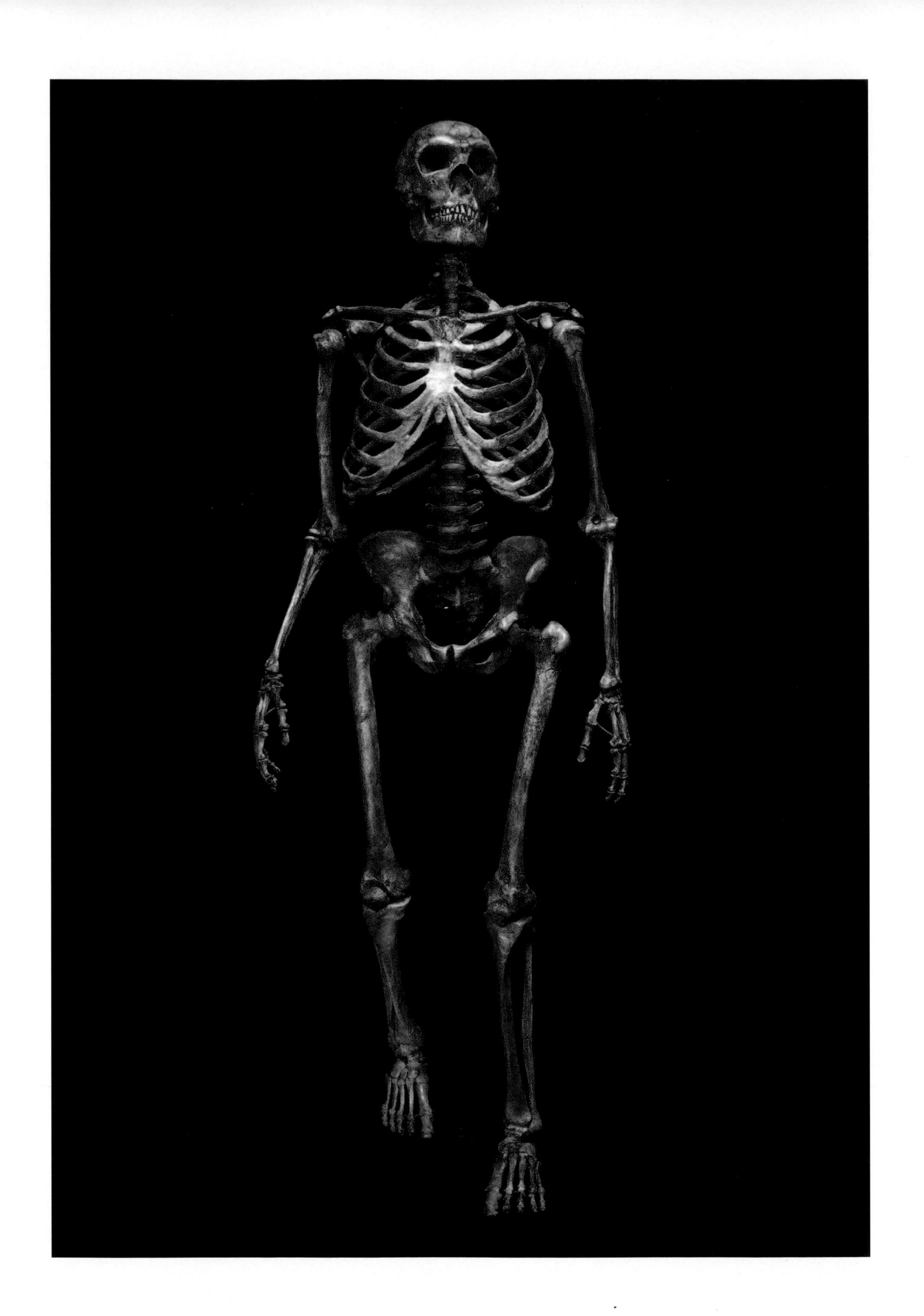

introduced considerable uncertainty in associating conditions at any one site with one or another event. Until absolute dating techniques were developed most Neanderthal dates were ballpark estimates. This is no longer the case. While most scientists still refer to glaciation events and animal assemblages as time markers, absolute dates for many Neanderthal finds now exist.

Biache-St.-Vaast, a river terrace site in northeastern France, provides the oldest evidence for Neanderthal occurrence. Absolute dates taken from stone tools found in the same layer as Neanderthal remains show that they lived here 175,000 years ago. Carbon isotope dating of Neanderthal remains from Vindija, a limestone cave 55 km north-northeast of Zagreb, Croatia, shows they occurred in Europe as recently as 27,000 to 32,000 years ago.

Tools and Culture

Neanderthals have been classically associated with a Mousterian tool culture, to the point that when only Mousterian tools are found Neanderthal presence at the site is assumed. Nonetheless, Mousterian tools have been found with non-Neanderthals (e.g., at Jebel Qafzeh) and Neanderthals have been found with Upper Paleolithic tools (e.g., St. Césaire with Chatelperronian tools).

Mousterian stone-toolmaking involves preparing a stone surface so that flakes of predictable size and shape will be produced when the stone is struck. The prepared stone or core is recognizable by its turtle shell–like appearance with numerous facets on one side resembling turtle shell segments. Once struck, the flakes are modified on one or both sides to produce the desired tools. Mousterian tool industry has the advantage over other industries in maximizing tool number made per unit of stone, and thus shows considerable forethought and skill.

Composite reconstruction of an adult male Neanderthal skeleton, using parts primarily from the La Ferrassie male, Kebara 2, and Feldhofer 1 partial skeletons.

Through Mousterian tool culture Neanderthals produced specialized tools for hunting, woodworking, and meat and skin preparation.

Neanderthals appear on occasion to have buried their dead. Positioning of corpses, use of burial markers, and the possible inclusion of flowers, pigments, and material objects in graves suggest death rituals. As evidenced in one individual from Shanidar who appears to have had multiple broken bones and a non-functioning arm, Neanderthal society may have cared for their sick and frail.

Considering their material culture, Neanderthals were probably capable of speech. Whether theirs was a fully symbolic language as complex as that of modern humans will probably never be known.

Appearance

Considering the high number of rainy days and associated cloud cover in some parts of their range, Neanderthals were probably fair skinned, with blue or hazel eyes, and blond or reddish hair being common. The face was big, with an extremely large and fleshy nose and most likely large ears. The temperatures in their habitats were probably not cold enough to select against facial hair (i.e., to cause frost bite as a result of water vapor in the breath forming ice on facial hair), and they most certainly would have had a beard.

Animals and Habitats

Nearly all animals that coexisted with Neanderthals are modern. Most small or medium sized mammals are still found today in the same areas that were once inhabited by Neanderthals. Most large animals, however, no longer exist in Europe, and a few are extinct. There were goats, bison (*Bison*), wild horses (*Equus*), wild pigs (*Sus*), wild cattle (*Bos*), and a variety of deer including the Irish elk (*Megaloceros*). In cold habitats, Neanderthal bearing deposits yield reindeer (*Rangifer*), musk ox (*Ovibos*), arctic fox (*Alopex*), and the extinct woolly rhinoceros (*Coelodont*) and woolly mammoth (*Mammuthus*). All the above were probably hunted and eaten by Neanderthals. Some Neanderthal sites yield cave lions (*Panthera*), hyenas (*Crocuata*), forest elephants (*Elephas*), and steppe rhinoceroses (*Dicerorrhinus*), animals that today are restricted to Africa or Asia. A large cave dwelling bear species closely related to grizzlies and brown bears are often found in caves inhabited by Neanderthals. Neanderthals carved and perforated the claws and teeth of these bears. Perforations suggest these carvings were hung or strung together and used as ornaments.

Neanderthals also shared their habitat with modern humans. Although modern humans are most often cited as the cause for European large mammal extinction, Neanderthals may have also played a part. Whether modern humans caused Neanderthal extinction through disease or competition, or Neanderthal populations were ultimately swamped by waves of immigrating modern human populations is unknown.

Climate

Considering their distribution, Neanderthals endured a wide range of climates, ranging from seasonally mild at their southern limits to nearly arctic at their northern and eastern limits. Haifa, a town 9 m above sea level and close to the three most southern Neanderthal sites (i.e., 30 km north of Kebara, 19 km north of Tabun, and 55 km west of Amud), records a relatively mild climate. January and February are the coldest months, with mean monthly temperatures of 13.9° C (57° F), and August the warmest month with a mean monthly temperature of 27.2° C (81° F). The lowest recorded temperature over a thirty-one-year period never dipped below freezing, and for 229 days out of the year temperatures never fell below 21° C (70° F). The climate is dry, with mean annual rainfall of 523 mm per year. Most rainfall occurs in November through January (64 percent; 335 mm). Over an eleven-year period, June-August did not record any rain.

Approximately one hundred kilometers west of the Neanderthal site of Shanidar, Mosul, Iraq, and 223 m above sea level records a climate similar to Haifa, but with somewhat more extreme temperatures and less annual rainfall (384 mm).

At Neanderthal eastern limits, Heyratan, a town 310 m above sea level and 125 km south of Teshik-Tash, shows monthly mean temperatures that never dip below freezing, but average lows for January that do. Over a twenty year period, October–March all recorded days with below freezing temperatures. The coldest and warmest monthly mean temperatures are in January (2.8° C; 37° F) and July (30° C; 86° F). With the Himalayas to the southeast and an expansive Asian continent to the north, weather may be extreme, with lows of –25° C (–13° F) and little rainfall. Over a twenty-eight-year period, mean annual rainfall was only 150 mm, with no precipitation during June-September. Most of the rain (110 mm; 73 percent) falls January–April and may do so in the form of snow. At 3100 m above sea level Teshik-Tash could be as much as 18° C (32° F) colder and may receive more precipitation, mostly as snow.

At their northern limits Dusseldorf, 40 m above sea level and 15 km west of the Neander Valley, records a temperate climate with moderate seasonal change. Coldest and warmest mean monthly temperatures are in January (1.7° C; 35° F) and August (18.3° C; 65° F). Mean monthly lows usually do not dip below freezing. Days with freezing temperatures, however, are recorded from October-April. An average 762 mm of annual precipitation is well spread throughout the year. Almost every month receives at least 56 mm of precipitation. March, with 38 mm, and April and October, with 46 mm each, are the exception. July receives the most rain (89 mm). The summer months (June-August) on average receive 30 percent more rain (244 mm) than the winter months (December–February; 185 mm). Ninety days out of the year it either rains or snows. By comparison other towns close to Neanderthal sites in Belgium and Hungary may record as much as 216 and 242 rainy or snowy days annually.

Because Neanderthals lived during the Wurm glaciation, their northern ranges would have had colder temperatures than they do today and possibly more precipitation. Expanding glaciers in the Alps and northern Europe may have exposed

Neanderthals year round to snow and ice. The glaciers must have been important factors limiting distribution of both Neanderthals and the animals and plants on which they depended. During glacial expansion, all animals, including the Neanderthals, would have been forced down to lower elevations and latitudes to find food. Nevertheless, latitudes at which Neanderthals lived would have insured relatively long days year round, warm temperatures in summer, and melting of snow and ice at the lowest elevations. Glaciers never reached southern Europe (i.e., the Iberian peninsula or southern Italy and Greece) and the Middle East. As such, the areas inhabited by Neanderthals were far from being as cold or dark as those inhabited by today's Inuits or Sami (Laplanders).

Classification

Neanderthals are the best-defined extinct human fossil species, and their remains are clearly recognized and identified as such. Differences between Neanderthals and modern humans (in both anatomy and genetics) are greater than those that exist between modern human populations, and equal or greater in magnitude to those existing between closely related living great ape species. Despite being contemporaneous with modern humans and appearing to have overlapping ranges at various points in their distribution (e.g., Middle East and later in central Europe), they appear to have maintained their distinctiveness. As such, a species designation (*Homo neanderthalensis*) separate from *Homo sapiens* is well justified.

Considering that evolution is a continuous process, but classification involves discontinuous leaps from one category to another, Neanderthal classification problems surface when following lineages back in time. Fossils from Sierra de Atapuerca and other European sites (e.g., Steinheim, Ehringsdorf, Reilingen Montmaurin, La Chaise Suard, etc.), which predate Neanderthals and may be presently referred *to Homo heidelbergensis*, are very similar to Neanderthals in both their skull and skeleton. In some cases, fragments of these fossils cannot be easily distinguished from Neanderthals. If *H. heidelbergensis* or subsets of this group turn out to be exclusive ancestors of Neanderthals, they should be referred to *Homo neanderthalensis*.

Historical Notes

The first Neanderthal, an infant with unerupted milk molars (about twenty months in modern human dental age), was found in 1829 by Charles Schmerling during excavations of Awir Cave II just north of Engis, near Liege, Belgium. Schmerling described this excavation and the skeletal remains he unearthed in a manuscript entitled *Recherches sur les ossemens fossiles découvert dans les cavernes de la province de Liège* (Liège, P.J. Collardin Editor, 1833-1834). Although aware the bones were truly ancient, Schmerling did not recognize them as a distinctive species. In March 1848, Lt. Flint reported the Forbe's Quarry Neanderthal find to the Gibraltar Scientific Society, but like Schmerling failed to recognize its full significance.

In August 1856, miners quarrying limestone in the Feldhoffer Grotto above the Neander Valley found the upper part of a skull vault, and fragments of an arm bone, ribs, thigh bones, and parts of a pelvis. Johann Carl Fuhlrott, a local schoolteacher, identified the fossils as those of an ancient human. In the *Quarterly Journal of Science* in 1864, William King made the Feldhoffer Grotto Neanderthal the holotype specimen for a new species, *Homo neanderthalensis*. The species name refers to the Neander Valley where the fossil was found. In that same year, George Busk reported the Gibraltar remains to the British Association for the Advancement of Sciences (published in 1865), recognizing them as similar to Feldhofer. Reports of La Naulette, Sipka, and Spy Neanderthal finds shortly followed in 1866, 1880 and 1886. Some time after 1885, Awir Cave, the earliest Neanderthal excavation, was destroyed by mining activity.

In 1882, Rudolf Virchow wrote that the Sipka Neanderthal was an abnormal modern human. In 1927, Alex Hrdlicka proposed Neanderthals as the evolutionary link between *Homo erectus* and modern humans. In 1936, Charles Fraipont finally showed the Engis 2 individual was a Neanderthal infant. Camille Arambourg, and then William Straus, and Alexander Cave, in 1955 and 1957, showed that the old man from La Chapelle-aux-Saints had arthritis, challenging the traditional view expounded by Marcellin Boule that Neanderthals walked stooped over with bent knees.

Based on DNA extracted from Neanderthals and modern humans, Matthias Krings and colleagues, in 1997 and 1999, estimated the date for the human–Neanderthal split to be four times greater than that estimated for the last common ancestor of all modern humans. The DNA findings together with a 27,000 to 32,000 year old date for late surviving Neanderthals, and the occurrence of modern humans in Africa at least 50,000 years earlier, put a final end to Hrdlicka's notion that Neanderthals were an evolutionary link between *Homo erectus* and modern humans. The holotype specimen of *Homo neanderthalensis* is housed in the Natural History Museum in Bonn, Germany.

This time he escaped the tall strangers with small faces. But how long could he run?

AND THEN

THERE WAS
ONE

HOMO SAPIENS

Skull, Teeth, and Diet

Modern human skulls are characterized by a relatively small vertically set face, a high forehead, and a large braincase. There are separate and modest bony brows above each eye, which when well-developed are most pronounced towards the midline, but separated by a depression from their counterpart on the opposite side. The lower jaw has a prominent and forward protruding chin. Generally, all teeth are relatively small and set in an arch. The front teeth are vertically set, and the canines do not protrude past the tooth chewing surface or show consistent size differences between males and females. Additionally, modern humans all share a complex of anatomical features associated with a vertically set face, a short, but wide skull base, and a unique lower jaw, neck, and head position.

Modern human skulls also show differences associated with geographic distribution that can be broadly used to identify skulls to one of five of the following geographic groups: 1) Caucasoids, Europe, North Africa and western Asia; 2) Mongoloids, eastern Asia and North and South America; 3) Blacks, sub-Saharan Africa; 4) !Kung San and Khoikhoi, Kalahari Desert and the Cape of Africa; and 5) Australoids, Australia, Tasmania, and New Guinea. Historically, geographic barriers (e.g., oceans, deserts, mountain ranges, rivers, etc.) separating these groups have reduced interbreeding between them and led to their differences. The effectiveness of these barriers to do so and the biologic reality of the human geographic groups are confirmed by the large number of mammals that show population discontinuities with a geographic pattern similar to those shown by humans. Since the end of the last Ice Age, however, the trend toward separate human populations has been reversed, and features distinguishing these populations increasingly blurred.

Modern human populations eat a wide variety of foods. For instance, Inuits eat high amounts of animal meat, but virtually no fruits, grains, or tubers. Other human populations, because of their convictions or religious beliefs, do not eat any meat at all, subsisting only on plant foods. Some cattle herding tribes along the Nile River eat relatively little meat, but drink milk and blood. Although populations living in developed countries consume large quantities of grain (e.g., rice, wheat, corn, and millet), hunting and gathering groups, such as the Congo forest pygmies and !Kung San, eat relatively very little grain. When compared to great apes, modern humans on average eat much more grain and more animal meat and less plant pith and leaves. Diet together with food preparation are probably the causes of relatively small molars, vertically set front teeth, and the shape and form of the human tongue and palate. Diet, and when and how we eat, has also had an evolutionary influence on many other modern human skull features, ranging from how wide we can open our mouths to our head position.

Skeleton, Gait, and Posture

The modern human skeleton is characterized by long lower limb bones, a short pelvis, short toe bones, and massive ankle bones. The chest is barrel-shaped, and the backbone increases progressively in girth from top to bottom. The four principal regions of the backbone—neck, trunk, lower back, and pelvis (sacral)—have curvatures that alternate from region to region, so that looking at the entire backbone from behind, the neck and lower back curve out at their ends (this curvature is referred to as lordosis) and the thorax and pelvis region curve in (this curvature is referred to as kyphosis). The tail, the fifth region of the backbone, is reduced in size. Its segments are fused and follow the curvature of the backbone's hip region.

The thigh bone's knee joint surface is set at an angle to the long axis of its shaft, forming a "carrying angle." This angle allows for column-like positioning of our shin bone so that it bears weight more effectively. Nearly all the human skeletal features that serve to distinguish us from other primates are associated with the manner in which we stand and move about on two legs. In one way or another, our unique features all serve to satisfy the mechanical requirements of supporting our body weight on just two legs with minimum energy expenditure, and maximizing the efficiency of our two-legged movements

Within humans, however, there is considerable variability in the skeleton, especially in the proportions of the different skeletal segments. For instance, sub-Saharan Africans are characterized by having proportionately long limb bones and finger bones, and Mongoloids (natives of eastern Asia) short ones. Conversely, Mongoloids have proportionately the longest body trunk lengths, while sub-Saharan Africans have proportionately the shortest ones. These differences can be linked to environmental differences in the geographic regions of origin. For example, a human with a short body and long fingers and limbs has a greater surface-area to volume ratio than one with a long body and short fingers and limbs, and thus loses body heat more quickly. The greater the surface area, the greater the rate at which body heat is lost to surrounding air.

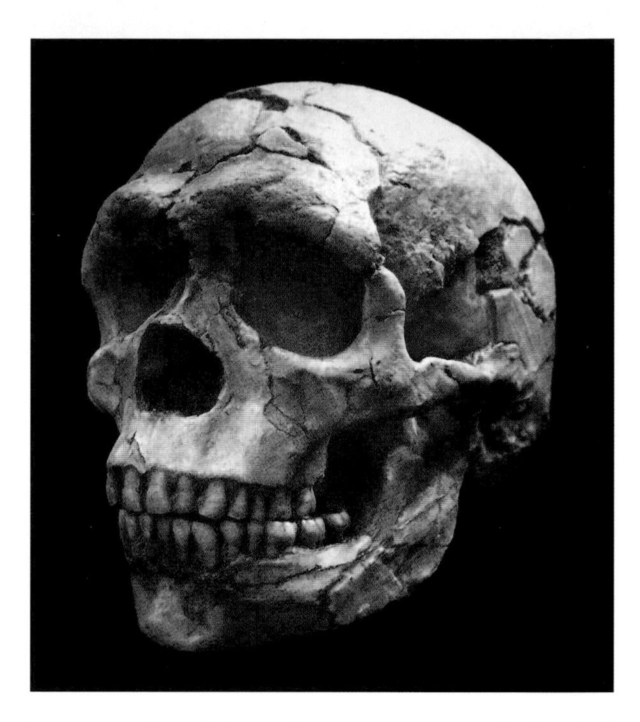

Reconstruction of the early *Homo sapiens* Qafzeh 6 skull with a modified Qafzeh 9 mandible (lower jaw).

Coming from cold areas, Mongoloids have reduced relative body surface area to preserve heat and stay warm. Coming from the warm climates of tropical and subtropical Africa, sub-Saharan Africans have increased relative surface area to prevent overheating. Although differences in body segment proportions may be associated with climatic differences, they also endow each of the geographic groups with different capabilities. For instance, the high lower-limb to body-size ratio makes sub-Saharan Africans much better natural runners than Mongoloids. Sub-Saharan Africans are thus much more commonly seen in Olympic finals in all the running events, despite the fact that Mongoloids currently make up a much higher proportion of the human population.

Given that differences in skeletal features may affect multiple functions, it is probably incorrect to interpret their evolutionary development as the solution to a single need. More often than not, these differences are solutions or compromises to multiple needs and have no simple adaptive explanation.

Fossil Sites and Range

Modern humans are presently the only known living members of the human lineage. They are found in fossil and sub-fossil deposits on all continents with the exception of Antarctica. Modern humans inhabit areas from below sea level to approximately five thousand meters above sea level, but may range up to the highest elevations found on Earth. Human populations, however, are most highly concentrated in temperate regions along rivers and coastal areas below 2000 m above sea level. Humans are much less concentrated in number in Arctic regions, deserts, and tropical forests, and at elevations above 3000 m. More than half of the 6.5 billion humans living worldwide live on the Asian continent. On average, three to four humans will die every second. A percentage of their remains will be fossilized or preserved and together with associated cultural and manufactured objects will produce a future record of their existence, distribution, and lifestyle.

Age

Modern humans encompass a very wide range of variation and as such there is no agreement as to what are "modern" humans, and the date they first appear in the fossil record. The earliest unarguably modern human remains have been found in the following localities, ordered according to geographic region. Africa: Dar Es Soltane Cave near Rabat, Morocco, 40,000 to 70,000 years; Hau Fteah Cave, near Benghazi, Libya, 47,000 years; and Klasies River Mouth Cave, south of Cape Town, South Africa, 60,000 to 80,000 years. Middle East: Jebel Qafzeh Cave, south of Nazareth, Israel, 80,000 to 94,000 years. Asia: Liujiang Cave, in Guanxi province, south central China, 67,000 years. Europe: in Czech Republic, Mladec Cave, in Moravia, 31,000 years, Dolni Vestonice, 26,400 to 29,000 years, Pavlov, south of Brno, 26,000 years, and Predmosti brickyard, northeastern Moravia, 26,000 years; in France, Abri Pataud, 20,000 to 32,000 years, and Combe-Capelle rock shelter, near Dordogne, 25,000 to 28,000 years; in Italy, Grimaldi Caves, up against the Monte Carlo border, 25,000 to 30,000 years. Australia: Mungo Lake, 20,000 to 40,000 years. North and South America: in USA, Marmes, 11,000 years, and Kennewick Man, Washington State, 8,500 years; in Mexico, Tepexpan, 11,000 years; in Peru, Guitarrero Cave, 12,500 years; and in Brazil, Lapa Vermelha, 11,500 years. Modern human evidence in the Americas is claimed to extend back to 40,000 years ago, but dates for older skeletal remains than those listed above were never documented (e.g., Midland Man, Texas, USA, 20,000 years). The earliest dates for modern humans, therefore, are in Africa and the Middle East, followed closely by Asia. Modern human appearance in Australia seems to occur before it does in Europe. The Americas are not populated until some time later.

In Africa there are a large number of human remains dating back to 200,000 years or more that are very much like modern humans with progressive modern human features (i.e., a slightly built face with a high forehead and a large braincase), but exhibiting some features not common to modern humans. Some of the "non-modern human" features, however, may exist at low frequencies in some modern human populations, e.g., a large single bony brow ridge in Australoids, the absence of a human chin in !Kung San, etc. As such, how inclusive a modern human classification is designed to be depends on whether it will encompass populations with a high frequency or fixed presence of these "variably non-modern human" features. When following more inclusive classifications, the earliest appearance of modern human is indisputably centered in Africa, as demonstrated by the geologic age of the following fossils: Jebel Ighoud, near Safi, Morocco, 130,000 to 190,000 years; Herto, east of Lake Yardi

Ominous clouds in the afternoon sky bring great worry to an early male *Homo sapiens* without shelter some one hundred fifty thousand years ago.

in the Middle Awash River, Ethiopia, 154,000 to 160,000 years; Omo Kibish I, Omo River Valley, along Nkalabong Mountain, Ethiopia, 104,000 to 196,000 years; Singa, along Blue Nile, Sudan, 133,000 years; Guomode, east shore Lake Turkana, Kenya, 180,000 to 270,000 years; and possibly Florisbad, Orange Free State, South Africa, 260,000 years.

Differences Between Males and Females

Human males are usually larger than females, showing mean body weights approximating a four to three ratio in most populations. Generally, males have joint surfaces which, relative to their bone shaft cross-section or length, are larger than those of females. Muscle attachment sites are often better developed than in females. The male lower jaw is more angular, and the hyoid bone, which sits above the voice box and anchors the tongue muscles, is also larger. Females have proportionately wider hips associated with childbearing. Differences in pelvis width and bony birth canal size are associated with a number of pelvic features that distinguish males from females. Unlike most other primates, there are no differences in canine size between the sexes. It is not clear how far back in our ancestry the modern human male and female skeletal differences developed, and how far back in our evolutionary lineage they still served to distinguish the two sexes apart.

Tools and Culture

The appearance of modern humans in Africa and Asia is usually associated with the presence of Late Stone Age tools. These are characterized by a greater proportion of carved bone, antler, and ivory, and also by stones that have been ground to desired shapes.

Deep in the caves of Altamera, a brother and sister Cro-Magnon breathe life into the image of a bison on the cave wall.

Modern human appearance in Europe has been classically linked with the presence of Upper Paleolithic tools and with a proliferation of art. The latter has been preserved in the form of cave paintings, carved objects, and stone blades and points that appear to sacrifice the utilitarian for the aesthetic. Upper Paleolithic art has been attributed to an improved capacity for knowledge and for symbolic reasoning in modern humans when compared to other members of the human lineage, specifically Neanderthals.

The earliest phase of the European Upper Paleolithic (i.e., the Chatelperronian), however, has been shown to be commonly associated with Neanderthals and not with modern humans. Modern humans thereafter were associated with the Aurignacian, a more recent phase of the Upper Paleolithic, which is distinguished from the Chatelperronian by an abundance of art including carved figurines. Since then deposits containing Aurignacian objects have been absolutely dated at 40,000 years, predating the first documented appearance of modern humans in Europe by approximately ten thousand years, and also predating the Chatelperronian.

Hypotheses that modern human appearance is associated with harvesting plants and small animals (especially marine-life), as opposed to big game hunting, need further testing. Nevertheless, there does appear to be a link between modern humans and exploitation of seashore resources, especially in Africa and Europe. Moreover, harvesting resources in great quantities at relatively little energetic expense is no doubt one of the most significant objectives of modern human culture.

Animals and Habitats

Modern humans share the planet Earth with all living animals. Because humans live in, range into, or in one way or another affect all of Earth's habitats, all of Earth's animals share their habitats with humans. There are, however, a number of animals that are known to flourish in human-made habitats. The black and the Norway rat, both with an origin in Asia, have spread with human trade in the last 1000 years to occupy all areas of dense human population. The domestic pigeon, a bird of rocky cliffs and ledges native to Eurasia, has flourished worldwide in man-made city-scapes. Like rats, pigeons are also considered pests. Some animals, however, have flourished because of their usefulness to humans. These include domestic farm animals (e.g., horses, cows, pigs, sheep, goats, and chickens), and house pets (e.g., cats and dogs).

A much wider variety of animals have flourished in man-made habitats in localized geographic areas according to the type of modifications that the habitat has undergone. For instance, in northeast United States, Canadian geese, benefiting from the well watered grass lawns and artificial ponds in suburban landscapes, have sharply increased in numbers. The mammoth garbage dumps and razed land (often paved) associated with dense human populations along the eastern United States seaboard have benefited seagulls, a marine scavenger. In Africa, marabou storks, black kites, and magpie crows have flourished as a result of an increase in man-made garbage in densely populated areas. As is expected, given the opportunistic nature of evolution, there are hundreds of more examples of animals benefiting from man-made habitats. Regardless of whether or not the modifications we make to Earth will ultimately lead to our extinction, some animals will benefit from these changes and life on Earth will surely go on.

Classification

All living humans are currently classified in the species *Homo sapiens*. As noted, this species can be broken down into five geographic groups. Because of present day human mobility made possible by jet planes, trains, automobiles, and ships, and the homogenization of human societies and culture, these groups over time are unlikely to further diverge. On the contrary, if the current worldwide culture persists, divergent geographic groups are likely to become more similar to each other. Those represented by the largest populations will probably swamp the others. Given current human mobility, however, the evolutionary significance of these geographic groups is not clear. Nevertheless, because each group shares a number of unique genes, both beneficial and detrimental, geographic groups have important implications for disease diagnosis and treatment.

Historical Notes

The name *Homo sapiens* for modern humans was first proposed in 1758 by Carl Linnaeus in the tenth edition of his *Systemae Naturae, per Regna Tria Naturae* (Stockholm: Laurentius Salvius). The name comes from the Greek "*Homo*," meaning "same," and Latin "*sapiens*," meaning "to know." In this same work, Linnaeus, based on a hypothetical concept, also named the species *Homo monstrosus* and *Homo troglodytes*. A binomial system of giving each organism a generic and species name had been introduced in earlier editions of Linnaeus' work. Since Linnaeus, there have been many proposed classifications. Most of these consistently recognize only one modern human species.

Exploration and colonization of new worlds, such as Mars, will be made possible in the near future by human ingenuity. Protected by a climate controlled exo-suit, a worker *Homo sapiens* proudly places the first earthling flag into a new planet's soil.

AFTERWORD

This book vividly brings to life twenty-two extinct species of our ancestors, branches of a once diverse human family tree. It shows the sequence of development of a combination of crucial adaptations that make us, the last surviving humans, the unique species that we are today. The first of these, bipedality, led our ancestors 6 million years ago to new environments and gave the potential for the developments that followed. Manual dexterity then provided the ability to manipulate a range of materials from large and heavy to small and delicate, and thus to develop technology 2.6 million years ago. Subsequently, encephalization (the trend toward larger brain size) resulted in increased intelligence and the potential for unprecedented innovations in our technology, the ability to remember and store information, to explore, to discover, and to invent. Finally our sophisticated communication, both oral and written, allows us to pass on knowledge and information to all other individuals of our species, and to future generations.

With these adaptations, we can achieve the impossible for all other species. We can see the world from outer space, and appreciate that we are all confined to one small planet, dependent on its resources for our survival. We can measure our population numbers and calculate that we now number over 6.5 billion individuals and that this number is increasing every minute. In a single generation, we have watched the volume of water, a vital resource, in many wells, boreholes, reservoirs, and rivers decrease dramatically; some of the largest rivers in the world now no longer flow to the sea. We can measure the acreage of forests that are destroyed daily, and we know that forest cover is disappearing at an unprecedented rate. We can see that we are irreversibly changing our planet and destroying the very systems that support us. And we can predict the course of these adverse trends and understand the urgency of taking actions to secure our future. Because of our increased numbers and wasteful utilization of resources, our activities are threatening the survival of our species and that of many others as never before.

Because we are intelligent, we know these things, and because we are intelligent, we have the ability to do something about it. We are unique in that we can plan for the future. We can educate our children, and they their children, to appreciate the urgency and necessity of taking serious action to stop the destruction of our planet. Through education and population control, these trends can be reversed. Inevitably, many millions of years from now, the Sun will become too hot for any life to survive on Earth, but for now we certainly have the ability to extend our brief tenure on this planet from one of the shortest to one of the longest.

This book puts humanity into perspective in time, in space, and as just one of millions of species that have inhabited Earth over the last 3.5 billion years. It has taken 6 to 7 million years for *Homo sapiens* to evolve from a common ancestor of modern apes to ourselves, and we have existed as a species for just 200,000 years. The success of a

BY MEAVE LEAKEY

species can be measured from several different angles. In terms of population numbers, we are undoubtedly successful, and increasingly so. In the last fifty years, our population numbers have increased exponentially from 2.5 billion to 6.5 billion. In terms of technological development, no other species has ever reached our level. We can probe outer space, penetrate the depths of the oceans, and accomplish undreamt of calculations through increasingly sophisticated computers. But in terms of longevity of our species, only time will tell. We are newcomers to this planet. From an evolutionary perspective, 200,000 years is just a blip, which raises the key question: Is our unique combination of adaptations, and most particularly our enlarged brain, a successful adaptive strategy in terms of the long-term survival of our species? Or will *Homo sapiens,* the only surviving species of a diverse evolutionary past, be a short-lived phenomenon that appeared and disappeared in a few hundred thousand years, inhabiting the planet for less time than the majority of our ancestors portrayed in this book?

THE SEARCH FOR FACES OF THE PAST

BY G.J. SAWYER AND VIKTOR DEAK

"The skeleton is the witness as to what the person was like in life. The bones speak to us. They speak softly. In fact, so softly that we cannot hear them, but we can see what they are trying to tell us."

— Clyde Snow (who identified thousands of victims of Guatemalan genocide from their skeletons)

Take an anatomist, a sculptor, a paleoanthropologist, a makeup artist, a forensic anthropologist, a painter, and a draftsman, and combine them all in one person. Then spark that individual with a vivid imagination, and add a burning obsession to create accurate images of human faces from the remote past—visages that no living human has ever seen. Put several such people together, and you have the Fossil Hominid Reconstruction and Research Team, the artist-scientists who created this book.

As humans, we have a self-awareness and curiosity about our own past that is unique among our fellow creatures. When Europeans and Americans first became aware of the existence of Egyptian mummies—the preserved bodies of individuals who lived some four thousand years ago—they flocked to museums to gaze at the shriveled, leathery faces of the pharaohs and pyramid-builders. As we view them laid out in glass cases, we instantly imagine how they must have appeared in life, in the streets or palaces of Thebes. But while these remarkable mummies continue to fascinate, they are not really very ancient. They are essentially ourselves. If we want to visualize the faces of humans and near-humans who lived 100,000 years ago—even a million or five million years ago—our best window into deep time is through the combination of art and science known as forensic anatomical reconstruction.

Forensic anatomical reconstruction is widely used today in missing-person and homicide investigations, as well as in physical anthropology, for visualizing the appearance of extinct human ancestors. Applied to beings that lived and died many thousands, or even millions, of years ago, the final result is like time-traveling to capture impossible portraits.

How do we go about conjuring up faces from the dawn of humanity? First, we obtain polyester or urethane resin cast reproductions of the best fossils from humankind's prehistoric past, many of which have been discovered during the past century—and some of the most important only within the past few decades. Upon these casts of ancient fossil skulls and bones, we carefully rebuild the soft tissue of the face, comparing its conformation and depth of tissue to those of living humans and apes. Our techniques, meticulously applied, can provide a three-dimensional picture of what our ancient relatives and ancestors probably looked like.

The earliest known attempts to artistically put flesh on bones for an exhibit are the elaborate anatomical sculptural works of Ercole Lelli of Bologna, Italy (1702-66). Lelli assembled and posed both human and animal skeletons, and meticulously sculpted the musculature over them in wax. So renowned were his accomplishments that they have survived to this day, and are still exhibited at the Anatomical Museum in Bologna, Italy, where they continue to elicit gasps of admiration from visitors. Many other artists and anatomists would follow, each contributing to and improving on the technique until it has become more science than art.

In the late nineteenth and early twentieth centuries, the American Museum of Natural History in New York City became a center for artistic reconstructions of extinct animals. (*See* Appendix 2.) Under the leadership of paleontologist Henry Fairfield Osborn, artists and sculptors were commissioned to create scientifically accurate images of living, breathing dinosaurs, wooly mammoths, and prehistoric humans.

One of the Museum's pioneers in reconstructing the faces of early man was the talented James Howard McGregor, who taught anatomy at Columbia University. A gifted sculptor as well as an anatomist and physical anthropologist, McGregor developed techniques that became the forerunners of those still in use today. Between 1914 and 1919, he created a series of bronze portrait busts of Neanderthal, early *Homo sapiens*, Java Man, and the (now discredited) Piltdown Man.

Using new tools, modern materials, and the remarkable array of more recent fossil discoveries, we have continued in McGregor's tradition—and so we have dedicated this book to him. Probably the best example of McGregor's work is his bust of the Neanderthal male from La Chapelle-Aux-Saints, which has delighted and inspired generations of young visitors to the Museum, including ourselves.

Various techniques can be used to reconstruct the face of an unknown individual. In recent years, using statistically "average" measurements of soft tissue depths, forensic investigators have programmed computers to rebuild faces over skulls for homicide and missing persons investigations. This method can quickly and efficiently work out a rough likeness, but has little value in recreating the faces of ancient human and pre-human ancestors. Fossil skulls are fragmentary at best, and there is only inferential reference data for comparison in reconstructing muscles and tissue.

By examining a modern human skull, a skilled physical anthropologist can usually determine its sex, possible "racial" affinities, age at death, nutritional status, disease, and various subtleties reflected in the facial bones, teeth, size and shape of jaw, and degree of muscular strength. But with ancient hominids, we usually cannot start with a complete skull as a base, and must first reconstruct one from available fragments—sometimes from several different individuals of the same genus and species. Only when we have a scientifically accurate reconstruction of the fossil skull can we attempt soft tissue restoration.

During the last century, artists, anatomists, and paleontologists developed two methods or "schools" of forensic reconstruction. The American school, pioneered in the nineteen forties by the anatomist Wilton M. Krogman, devised a technique in which individual tissue depth markers are plotted over the facial part of the skull. These markers or points are then connected by a network of clay strips. This lattice of clay strips, connecting individual tissue depth markers or points, is then filled in or blended into a smooth skin surface.

The result is a mannequin or "generic" human face. Though practical for some purposes, the technique gives crude results because it is based on average or "standard" tissue depths. It ignores fine details about the skull's individuality, geometry,

The Fossil Human Reconstruction and Research Team at the American Museum of Natural History. Left to right: G. J. Sawyer and Viktor Deak.

muscular scar topography, and overall morphology, and therefore masks many clues. Such images are most useful in missing persons cases, where a more general description will draw a larger number of interested parties to make inquiries and view the model. We cannot use it, however, to accurately reconstruct faces of very pre- and ancient humans, which are different from modern skulls in many important features.

The other basic technique, known as the "Russian method," was developed by the Russian anatomist and paleontologist Mikhail Gerasimov (1907-70). His method requires greater, time-consuming precision, but allows much more fine-tuning. Individual soft tissue anatomy (muscle layers, glands, and fibro-fatty tissue) is meticulously rebuilt in clay, using tissue depth measurements where applicable. Eighteen basic facial muscles are sculpted in place. The technique is best described as a "dissection in reverse." The reconstructor uses the skull not only as an armature, but as a detailed anatomical guide in the soft tissue placement. He or she is able to utilize the skull's peculiarities, asymmetries, pathologies, sexual dimorphic changes, and muscular attachments, allowing not merely a generic likeness, but a portrait of a particular individual.

Gerasimov's Russian method recreates the nose by extending one line from the bony bridge and another from the floor of the nasal opening, and then rounding off their point of intersection at the tip. The nostrils are on average 1.67 times the width of the bony nasal opening in the skull. Corners of the mouth lie directly below the inner edges of the irises in the eyes, or between the lateral edges of the canine teeth, which are usually about the same width.

Our reconstruction technique is based on the Russian method. We carefully build up clay models of the missing facial musculature, glands, fibrofatty tissues, depth measurements (where applicable), and skin from the skull outwards. Starting with the deepest individual musculature tissues, the face is slowly created, layer by layer according to what the geometry, size, and muscular landmarks of the skull tell us.

With the more recent skulls, we have better comparative data than did our predecessors. Where McGregor's measurements, for instance, were based on cadavers, the ones we use today are taken from living people using ultrasonic echo location and CT scans, and are therefore more reliable. Of course, tissue depth measurements derived from living (extant) populations can be used only as a rough guide for extinct ones. Both McGregor and Gerasimov used anatomical technique as well as tissue depth measurements for their Neanderthals—the same combination of methods that we still use today.

Today, however, we have a wide range of sophisticated materials to help us create vividly lifelike heads. To begin, we use high quality, low shrinkage silicone rubber to make a negative mold of the skull, lower jaw, and any associated upper torso skeletal elements. From this mold we produce a positive cast in a fast setting urethane plastic that faithfully reproduces every minute detail.

Of course, paleoanthropologists rarely find complete fossil skulls, let alone full skeletons. At best we have to work with only parts of a skull, jaws, teeth, and some limb elements. If the skull or skeleton (as in the case of our Neanderthal reconstructed skeleton) is incomplete, then we must mold and cast parts from other similar specimens, then combine them to create a complete composite.

In the case of the reconstructed Neanderthal skeleton, we used the main body parts from two nearly complete individuals, representative parts from a third, and many fragments from several others. We incorporated all the recovered parts of a single adult male from the site of La Ferrassie, France. In assembling this composite, we produced the first complete Neanderthal skeleton not only ever seen. Our reconstructed Neanderthal skeleton has not only extended our knowledge of what one of these individuals looked like, but enables us to better understand the bio-mechanical functions of the bony anatomy in walking and running. It further gives us a comparison between other fossil hominid and modern body proportions.

Faces, however, are our main concern, and it is the skull that dictates where and how the soft tissue anatomy is built up. The structure of the australopithecine facial skeleton suggests that the lips and nose were similar to living apes and not at all human in form, yet the small canines and rather human-like teeth suggest a face not quite fully pongid but different from all known living apes. As we progress, however, to more human-like forms like *Homo habilis* or *Homo rudolfensis*, we are confronted with a combination or mosaic of australopithecine (ape-like) and human morphologies. So we must adjust our reconstructions according to what the bones are telling us to do.

Finally, as we move on to such true humans as *Homo ergaster, Homo erectus*, and Neanderthal, we are at last able to draw more and more upon our knowledge of modern populations. Living humans possess a unique soft tissue anatomy in the middle or cheek region of the face called the superficial muscular aponeurotic system—a mesh of nerves, blood vessels, and fibro-fatty tissue that connects the underlying facial musculature to the skin. This mesh of tissue and fat fills out the face. If we sculpt only the muscles and glands on the skull of *H. ergaster*, we can produce a face, but it will have a very gaunt or sunken look. When we add the superficial muscular aponeurotic system, and refer to modern tissue depth measurements as guides, the face fills out and appears much more lifelike. It is here, therefore, that we are able to use an interface between what we know of modern human anatomy and apply it to a prehistoric skull with reasonably credible results.

The penultimate touches are the skin and glass eyes, which are positioned in the center of the eye sockets. The mold is then filled with a urethane rubber that is tinted with a flesh color, which of course is the artist's best guess. Finally, more color is judiciously applied to cheeks and other areas, and finally head, facial, and body hairs are punched in one by one, using a needle to insert them into the flesh-like rubber. The entire process, although very laborious and time consuming, is ultimately worthwhile if at the end we can gaze into a face that hasn't been seen for half a million years or more.

In recent years, digital cameras and computers have become indispensable to our work. Because casts of newly discovered fossils are unavailable until they are fully described and published—which can take years—sometimes we must proceed without them. In some cases, even when casts are available, they are too fragmentary upon which to base three-dimensional reconstructions. Some of the recent finds, such as *Orrorin tugenensis* and *Ardipithecus ramidus*, are known from either very incomplete skulls, or no skulls at all. In the absence of good skull casts, we have turned to digital imaging techniques.

Our reconstruction of the Dminisi hominid, which has been designated *Homo georgicus*, for example, was created entirely in cyberspace. By making photographs of three of our closely related hominid reconstructions, we established a virtual database for lighting, visual textures, and background. Using several digital photographic views of one of the more complete *H. georgicus* skulls, we calculated the depth of anatomical markers, then built up the muscles over the virtual skull in the computer image. Finally, using our database compiled from similar reconstructions, we added eyes, hair, skin color, wrinkles, and expression to the image. Executed with meticulous care, the results are virtually indistinguishable from physical reconstructions made by applying clay and rubber to a cast of the skull.

Physical anthropologist Ian Tattersall of the American Museum of Natural History has been our chief scientific mentor through this long and arduous project. In the beginning, Ian was a bit skeptical about the validity of various parts of the process. As he watched and guided the reconstructions, however, he became increasingly enthusiastic about the accuracy of our methods, and has been a steady source of guidance, advice, and encouragement.

How accurate are these reconstructions? In the end, they are only as good as the skill, experience, and education of the team that produced them. At every stage of the work, however, we invited the supervision of other physical anthropologists and paleontologists, who were familiar with the specific fossil hominids we were attempting to reconstruct. Some of our advisors were scientists, who were involved in the actual fossil discoveries, or had studied first hand the original fossils, and were therefore closely familiar with the individual hominids. So these reconstructions are not the unbridled fantasies of artists, but rather the result of an exacting creative interface between art and science. We've tried our best, in the words of the classic Barbra Streisand song, to retrieve from the past "pictures...of the way we were."

PORTRAITS OF PREHISTORY

BY RICHARD MILNER

IMAGING OUR ANCESTORS

Judging from their astonishing paintings and engraved images of animals on European cave walls—works that have somehow survived since prehistoric times—people have been making pictures for at least thirty millennia, and probably for a lot longer. In contrast, attempts by artists and scientists of our own day to make credible likenesses of the cave painters and their more remote evolutionary antecedents go back only a puny 150 years. Despite its short history, however, the vigorous post-Darwinian genre known as "paleoart" has had an enormous impact on our collective imagination.

As the latest fruit of that tradition, this book offers a new portrait gallery of our ancient ancestors and relatives, the outcome of a creative collaboration between physical anthropologist Gary Sawyer and paleoartist Viktor Deak.

Both men independently shared a childhood obsession with prehistoric humans. Sawyer, a New Jersey native, was inspired by W. M. Reed's children's classic *The Earth for Sam* and, later, by Charles R. Knight's classic murals of dinosaurs, mammoths, and cavemen at the American Museum of Natural History. Deak grew up in a leafy, suburban Connecticut town that may seem an unlikely place to dream about living the life of Neanderthals. In 1991, however, at age fourteen, he saw a *National Geographic* television show in which the paleoartist John Gurche sculpted his reconstruction of *Australopithecus afarensis.* "I was bitten by the bug," he recalls. "I knew immediately that I wanted to do what he did."

Deak's grandfather fueled his growing interest by giving him Zdeněk Burian's book of paintings, *Prehistoric Man.* Viktor told me he feels a personal connection to the images he creates: "I see myself in these people, living thousands of years ago. I'm haunted by going back in time."

As young Viktor sketched fantasies of the remote past, he did not yet realize that he would need a scientific collaborator to discipline and focus his talents. When he was twenty-six, he met Gary Sawyer at the American Museum, who was looking for an artist to work with him on reconstructions of early humans. And the rest is, well, prehistory.

Together they have joined the handful of bold, devoted souls who have dedicated their lives and talents to bringing dry fossil bones to life. Although the subjects of their attentions go back millions of years, the genre of paleoart is a surprisingly recent tradition. Artists had long depicted imaginative views of humans in a primeval world, but their reference point was Scriptural, not scientific. Remains of early humans began coming to light only two hundred years ago, and most other hominid remains have been discovered only during the past century.

One of the earliest archeological reports of evidence for early humans was that of the English antiquarian John Frere, who in 1800 published his ***Account of Flint Weapons Discovered at Hoxne in Suffolk.*** Workmen digging clay for bricks had come across finely worked flint hand axes in a layer of gravelly soil, sealed beneath a sandy layer containing mammoth bones. Frere concluded that the tools

[FACING PAGE] During the 1870s, Charles Darwin's protégé Sir John Lubbock commissioned Ernest Griset's watercolor *The Bison Hunt,* (above)—one of the first post-Darwinian works of paleoart. In this painting and its companion piece, *The Mammoth Hunt,* (below), Griset created a classic triangular composition with the hunters' spears angled up at the animals' head. (Courtesy of and © Bromley Museum, Orpington)

A fifteenth-century stone "wodewose" or "wild man of the woods" atop St. Mary's church in Mendlesham, England. His wild hair, beard, and ever-present club are iconic. Photograph courtesy of and © by John Telford-Taylor.

Pithecanthropus alalus, meaning "apeman without speech," was named by German zoologist Ernst Haeckel long before a fossil "apeman" was ever found. Haeckel's theories inspired this 1887 etching by Henri de Cleuziou.

Ernest Griset.

were "fabricated and used by a people who had not the use of metals. [They lived in] a very remote period indeed; even beyond that of the present world." (Incidentally, John Frere's great-great-great-granddaughter was Mary Leakey, wife of Louis Leakey and mother of Richard Erskine Frere Leakey, all of them distinguished paleoanthropologists.)

Although Frere's discovery went unnoticed until long after his death, further evidence of early humans continued to accumulate. Following the lead of the French prehistorian Jacques Boucher de Perthes, who trained workmen to search for stone hand axes in the 1840s, others began to seek and find quantities of prehistoric stone tools all over Europe. Part of a fossilized Neanderthal skull was discovered in a cave in the Neander Valley, near Düsseldorf, Germany, in 1856, a find that brought the term "caveman" into popular culture.

Beginning in 1858, when rich prehistoric deposits were discovered at Brixham Cave at Torquay, in Devon, England, the archaeologist William Pengelly developed revolutionary new techniques for conducting excavations. His systematic work at Brixham and nearby Kent's Cavern over the next two decades yielded tens of thousands of fossil animal bones and early human artifacts, and established their association in time.

English naturalist Charles Darwin's book *On the Origin of Species* shook the world in 1859 with its one-two punch: evolution by natural selection and the immensity of geologic time. Its impact was seismic, but even before it appeared, archeological discoveries that humans had lived with extinct mammoths and rhinos in Britain had caused many to question traditional beliefs about human origins. In 1851, Victorian art critic John Ruskin had lamented in a letter to a friend that his trust in biblical authority was being daily eroded by "those dreadful [geologists'] hammers." "I hear the clink of them at the end of every cadence of the Bible verses," he wrote. Now cavemen began to challenge Adam and Eve as primal ancestors in the popular imagination.

It turned out that some of the ancient "cavemen" were fine artists. In 1879, the first-known painted cave was accidentally discovered at Altamira, Spain; its images of extinct aurochs, bison, and horses stunned both the art and scientific worlds. Only rarely in subsequent finds, however, had the ancient artists portrayed themselves, and never with the sophisticated realism they had applied to animals. That state of affairs cried out for modern artists to reconstruct the appearance of what became an expanding roster of extinct humans and near-humans. The nascent genre of paleoart, which had taken hold in the 1840s to visualize dinosaurs and other fossil animals, expanded to portray extinct humans as well. (The term "paleoart" to describe it, however, was not coined until much later; its origination is often credited to painter Mark Hallett, who began using it to describe his own work in the 1990s.)

John Lubbock, Darwin's informal (and only) student, commissioned some of the first paintings in the genre. The scion of a banking family that owned much of the Kentish countryside surrounding Darwin's home, Lubbock decorated his indulgent father's mansion, High Elms, with a collection of primitive stone tools, ethnographic artifacts, glass-enclosed colonies of social insects, and eighteen watercolor paintings of early humans going about their daily lives. The paintings, which Lubbock sponsored during the 1870s, were the work of Ernest Griset, an outstanding natural history illustrator whose anthropomorphic animal drawings often lent whimsy to the pages of the magazine *Punch*. Lubbock himself had coined the terms Paleolithic and Neolithic, meaning old and new stone ages respectively, in his landmark book, *Pre-Historic Times,* which appeared in 1865. The book also includes the earliest printed usage of the phrase "cave-man."

Neither Lubbock nor Darwin, however, published any of Griset's pictures in their books. All these paintings, which were done during the 1870s, now reside in the Bromley Museum in

Orpington, England, where some of them are featured in a permanent exhibition about Lubbock and his collections.

Since only a few fragmentary fossils of early man had been found by the 1870s, the Griset-Lubbock collaboration did not attempt to define their subjects' faces. Griset painted "generic" prehistoric people, viewed from a distance hunting mammoths and bison, living in tent villages, or making dugout canoes. Sequestered for a century by the Lubbock family, they were rarely seen, yet are important because of Lubbock's close association with Darwin. In his own ***Descent of Man*** (1871), Darwin refrained from including any artistic reconstructions, although he speculated that human precursors originated in Africa and were originally hairy, with large teeth and pointy ears.

This romantically idealized pair of Paleolithic "noble savages" resembles Adam and Eve transplanted from a biblical tropical garden to an Ice Age cave. An etching by Emile Bayard from Louis Figuier's book *L'homme privitif*, 1870.

Thanks to the generosity of the Bromley Museum authorities, two of these seminal paintings, ***The Mammoth Hunt*** and ***The Bison Hunt***, are published here for the first time. In Griset's dramatic rendering, the humans appear puny and vulnerable—not at all masters of their environment, but like a pack of wolves, of which several are trampled and killed while hunting these ferocious behemoths.

Many nineteenth-century artists continued to depict Adam and Eve as our earliest ancestors, reflecting the church-approved tradition. (Typically, they were shown without navels, reminding viewers that the biblical characters were not born of woman.) By Victorian times, however, some painters and illustrators began to conflate our Ice Age forebears with the traditional primal pair, their semitropical Garden traded for caves and frozen landscapes—but still uncorrupted by the vices and deceits of "civilized society." These were the "noble savages," whose praises had been sung by the eighteenth century philosopher Jean Jacques Rousseau. Yet, stooped, hairy "cave people" were sometimes depicted as Adam and Eve's "degenerate" descendants, who had fallen from Grace and were "dead ends" rather than precursors of humanity.

Beginning in the 1880s, astonishing discoveries of European caves with sophisticated Paleolithic paintings of long-extinct animals fired the imaginations of scientists and the general public alike. Still, many people continued to imagine our Ice Age ancestors mostly as wild-haired, club-toting clods. They were apparently not troubled by the evidence that the earliest known Paleolithic stone figurines depict women with elaborate coiffures, and that many cave paintings are extremely sophisticated—not at all childlike or "primitive."

Perhaps that notion of the murderous hairy brute was carried over from a medieval tradition of "wild men of the woods" or "wodewoses" in European art. Anthropologist Judith Berman has suggested that these archaic 15th-century iconic figures, with their unkempt hair and ever-present clubs, prefigured and continue to influence our images of the caveman, even today. Statues of the bizarre and intriguing wodewoses can still be seen in some of England's older rural village churches, particularly in Suffolk.

Some scientists commissioned artists to illustrate the presumed "missing links" between apes and humans. Ernst Haeckel (1834–1919), the German zoologist who popularized evolution in continental Europe, postulated the existence of *Pithecanthropus alalus* (literally, "ape man without speech") years before any fossils resembling an "ape man" had been found. He published drawings of the conjectural "reconstruction," complete with traditional club, this time with a hand-axe hafted to the end of it. When the Dutch physician Eugene Dubois, inspired by Haeckel, sought and found a heavy-browed skullcap of the Java "ape-man" in 1891, he named it *Pithecanthropus erectus* (now known as *Homo erectus*). Haeckel telegraphed him, "Congratulations to the discoverer of *Pithecanthropus* from his inventor."

Shortly after that episode, an indignant international committee of zoologists decreed that scientists could no longer name new species before they were actually found. (Haeckel had also proposed that another early hominid remained to be discovered, which he named *Homo stupidus*; I think it's safe to presume that it never went extinct.)

In 1907 zoologist Alexander F. Kohts and his wife, the biologist Nadezhda N. Ladygina Kohts, founded Moscow's State Darwin Museum, the

Just prior to the First World War, Belgian sculptor Louis Mascré, working with anthropologist Aimé Rutot, created the first portrait gallery of prehistoric humans. While their Neanderthal (left) looks entirely human, they thought he was dim-witted. In his book *Prehistoric Man* (1960), Czech paleoartist Zdeněk Burian depicts a very different-looking Neanderthal (right), one with a human-like face on an apish body.

world's first museum dedicated to evolution. Karl Marx and Friedrich Engels had thought Darwin an appropriate hero for a secular state dedicated to human progress. Kohts and his staff sculptor, Basil Watagin (1918–1948), set about creating one of the earliest halls of human evolution in a major museum. It contained no fossils, but exhibited an imposing and magnificent display of reconstructive statues and paintings of early hominids and prehistoric humans.

Just before World War I, a gifted Belgian sculptor named Louis Mascré and his scientific mentor, Aimé Rutot, completed their ambitious five-year project of creating a sculpture gallery of fifteen realistic prehistoric portraits for the Belgian Royal Academy of Sciences in Brussels, where they still remain on display. The two collaborators were aiming not only to reconstruct faces from fossil skull fragments, but also to imbue them with appropriate cultural attitudes and behavior. In depicting the relationship between Cro-Magnons and Neanderthals, for instance, they speculated that the "higher," more intelligent Cro-Magnons had enslaved the fierce-looking but dull and submissive Neanderthals. As Belgians, whose nation at that time was engaged in subjugating and exploiting Congo peoples, Mascré and Rutot may have found this a not unfamiliar arrangement.

During the 1870s and 1880s, the French artist Emmanuel Fremiet (1824-1910), already famous for his charming sculptures of animals, turned his hand to painting prehistoric scenes. The pictures proved popular in the Paris Salon, but, since he was not an evolutionist, his depictions of early humans looked physically just like modern Parisians. Then, inexplicably joking that his sketch of an ape showed "the exact portrait of Adam," he made a nightmarish sculpture of a powerful, gigantic gorilla carrying off a terrified nude woman, which won him a medal at the Paris Salon of 1887 and was the centerpiece of the show—anticipating the public's fascination with *King Kong* by almost fifty years.

The undisputed king of the paleoartists, however, was Charles R. Knight (1874–1953), who inspired all who came after him. The imperious paleontologist Henry Fairfield Osborn, president of the American Museum of Natural History from 1908 until 1933, hired the gifted young painter and teamed him with the museum's best anatomists and paleontologists. Together the teams created the most accurate and realistic reconstructions of ancient animals and early humans and near-humans that had ever been attempted.

Under Osborn's direction, Knight also executed a very influential series of paintings of early humans and near-humans for the museum's Hall of The Age of Man. Osborn's theories of what early men looked like and how they acted were based in equal parts on fossil skulls and on his own ideas about "aristogenesis," the "evolution of the best." For Osborn, the sweep of human evolution justified the elevation and celebration of his own social class: the "racial superiority" of so-called Nordic peoples. One contemporary critic attacked him (and, unfortunately, the entire Darwinian theory of evolution) by charging, "With enough good artists, plaster of Paris, and paint, Osborn can prove anything."

Paleontologist Stephen Jay Gould considered Charles R. Knight "more influential in shaping the public's understanding of dinosaurs and extinct creatures than any professional paleontologist who ever lived," although Knight remained almost invisible in scholarly journals. Gould wrote in his 1989 book *Wonderful Life*, "Not since the Lord himself showed his stuff to Ezekiel in the valley of dry bones had anyone shown such grace and skill in the reconstruction of animals from disarticulated skeletons."

Certainly, Knight's pioneering work had a profound influence on his contemporary, the anatomist and sculptor James H. McGregor, and such subsequent paleoartists as Zdeněk Burian, John Gurche, Jay Matternes, and (through them) Viktor Deak. Indeed, Knight's *Tyrannosaurs*, *Brontosaurs*," and cavemen (and even his mistakes, based on fragmentary skeletons) were endlessly copied and

[FACING PAGE] In this mural-size 1925 oil painting (above), American Museum of Natural History paleoartist Charles R. Knight pays homage to Cro-Magnon artists of 20,000 years ago. Knight based his own reconstructions of Ice Age mammals on the cave artists' renditions. In the 1996 Jay Matternes acrylic painting (below), a group of *Homo habilis* prepares to feast on a zebra carcass on the African savannah. Some members of the group drive off wild dogs that made the kill, while others chip stone tools with which to cut up the meat. (Top © American Museum of Natural History; bottom Courtesy of and © Jay Matternes)

Sculptor Basil Watagin poses among his creations in the Hall of Early Man in Moscow's Darwin State Museum, around 1921. The exhibition featured many paintings and sculptures of human prehistory, but contained no original hominid fossils. (Composite of Moscow museum © Richard Milner)

recopied not only by museums and textbooks, but also by movies, comic books, and toy-makers the world over. As each new Knight mural of mammoths, dinosaurs, and cavemen first went on display at the American Museum, an eager public formed long lines to see them.

During the spring of 1927, Knight visited the French painted caves to see the Ice Age artists' renderings first hand. He wrote of gazing at the mammoth and reindeer images "with unalloyed pleasure and astonishment" and with "a distinct feeling of awe and admiration for the skill of the man who had painted [them] thousands of years ago."

Knight expressed his identification with Paleolithic artists in his often reproduced 1924 mural of Cro-Magnon painters. One individual who is sketching a mammoth on the cave wall seems to have the face of Knight himself. If you add some steel-rimmed glasses to photos of the original, I believe the resemblance would be striking. Unfortunately, that likeness did not survive a repair

Originally drawn for the British humor magazine *Punch*, the *Prehistoric Peeps* series by Edward Tennyson Reed was wildly popular. In this 1895 cartoon, prehistoric cricketers are using the Stonehenge arches as wickets for their game, which is disastrously disrupted by a hungry reptile of indeterminate species.

job on the damaged mural in 1993, when the restorer ignored my entreaties and painted in a very different face.

When attempting to bring wooly mammoths back to life for his museum murals, Knight had been stumped on precisely how to reconstruct the living animals from their mounted skeletons. After viewing the French caves, however, he decided to rely on the truthfulness of the caveman artists. All of their engravings and paintings of mammoths are in precise agreement with one another on the high hump of fat on the forehead, the notch between the hump and the neck, and other anatomical details. So Knight decided to take art lessons from the (truly) Old Masters, and his own paintings defined those animals for all subsequent paleoartists.

Knight in turn inspired one of the greatest—and most prolific—of all painters of prehistoric life and early humans, the twentieth-century Czechoslovak artist Zdeněk Burian (1905–1981). One of the greatest—and most prolific—of all painters of prehistoric life, Burian left a magnificent legacy of more than 15,000 artworks and many more sketches and preparatory studies. Fascinated with natural history since his boyhood and directly inspired by the work of Charles R. Knight, Burian turned his remarkable abilities to depicting the lost world of prehistory.

A virtuoso, he could complete a painting in a few days that other artists might labor over for months. Almost immediately, his ability to imbue extinct creatures with vitality, as well as the dramatic tension of his compositions, began to win international recognition.

In 1932, paleontologist Josef Augusta saw some of Burian's illustrations in an early book, *The Hunters of Mammoth and Reindeer,* and sought out the painter. Together, they produced the most beautiful series of popular literature on prehistoric subjects created up to that time, including *Prehistoric Animals* (1957), *Prehistoric Man* (1960), *The Book of Mammoths* (1963), *Prehistoric Sea Monsters* (1964), and *The Age of Monsters* (1966). With another scientific collaborator, he produced three more profusely illustrated volumes on the life of prehistoric humans and near-humans.

In Czechoslovakia, Burian's paintings have been declared national treasures which by law cannot be sold or exported from the country. The government built a small Burian Museum to house a permanent exhibition of sixty of his works, a tiny proportion of his prodigious output.

As more European and American artist-scientist teams began to turn out ever more sophisticated reconstructions of early hominids, various tongue-in-cheek takeoffs immediately followed. Spoofs and cartoons of the caveman image appeared from the 1880s on, and some became extremely popular.

In 1893, Edward Tennyson Reed (1860–1933), who began by caricaturing prominent politicians for the British humor magazine *Punch*, launched a delightful series of cartoons called *Prehistoric Peeps* in that British humor magazine. One of his best known drawings shows a tribe of cavemen playing a form of primitive cricket at Stonehenge, using the giant stone arches as wickets. A decade later, a hoaxer planted a mammoth bone carved into a "cricket bat" at Piltdown, Sussex, to go along with a forged fossil skull—The "Earliest Englishman" with the earliest cricket bat. British scientists didn't get the joke until forty years later.

Three Ages (1923), a silent film classic, starred Buster Keaton (who also directed) as a prehistoric "average guy" trying to win his lady fair. Buster is thwarted by a monstrous, club-wielding pop archetype of a Paleolithic caveman. (Courtesy of and © Motion Picture Academy of Arts and Sciences)

In 1905, a Hollywood live-action silent comedy called *Prehistoric Peeps*, based on Reed's work, became the first dinosaur film, antedating the silent version of Sir Arthur Conan Doyle's adventure *The Lost World* (usually considered the first dinosaur movie) by two decades. In *Peeps*, the cavemen and dinosaurs were shown living together, a tradition that persisted through the comic strips *Alley Oop* and the *Flintstones* of the 1970s—and perhaps fueling later Creationist fantasies of dinos living with prehistoric people.

The Dinosaur and the Missing Link (1915) was a goofy comedy about two cavemen battling for the hand of Miss Araminta Rockface, featuring early stop-motion dinosaurs by Willis O'Brien. A genius at the film genre he pioneered, O'Brien eventually created the masterpiece of stop-motion animation, *King Kong,* in 1933.

Reed's drawings of dinosaurs hitched to carts ("the first Hansom cab") drove the British paleontologist Reverend E. A. Hutchinson up a wall. Annoyed by the dissemination of such anachronisms, the humorless cleric produced a book of cartoons called *Primeval Scenes* (1899) in which he eliminated dinosaurs and sea serpents from the caveman's world so as "not to mislead young people," and substituted "more accurate" Pleistocene mammals. The old pedant should have stuck to writing science books; without the dinos, his cavemen cartoons simply were not funny.

Part of the humor in caveman cartoons like Reed's "The First Golf Game" or "Early Cricket" is that they self-satirize the obvious projections of our own culture onto the caveman's world. Their "prehistoric" mother-in-law or argumentative spouse jokes iterate the eternal universality of the human condition. Caveman cartoons were also perfect vehicles for absurd anachronisms. Contemporary cartoonist John Callahan, for instance, shows a caveman watching television, and the screen cautions "the following program contains language."

Cartoonist F. Opper (1857–1937), in *Our Antediluvian Ancestors* (circa 1903), a collection of his drawings first published in the *New York Evening Journal,* opined, "although everything else in the world changes constantly, Human Nature has not changed, is not changing, and will never change." In 1923, Buster Keaton, a comic genius of silent films, appeared as a lovelorn caveman in *Three Ages*, in which he pursues fair lady through the Stone Age, ancient Rome, and the 1920s. Buster's Paleolithic rival is a club-wielding, wild haired, bearded giant—a classic pop caveman.

Despite sporadic, misguided attempts at scientific accuracy in prehistoric comedies, cavemen and dinosaurs eventually got back together in cartoons and movies. In 1935, for instance, Sunday newspapers introduced a leopard skin-wearing "cave boy" called Peter Piltdown who was always running from dinosaurs—an eponymous comic strip by Mal Eaton that lasted for more than three decades. When Piltdown man was exposed as a fossil forgery in the 1950s, the cartoon character survived under a new name which readers of *Boy's Life* magazine still remember: Rocky Stoneaxe.

Later popular comic strips featured time-traveling caveman Alley Oop, who rode his trusty saurian Dinny, and Fred Flintstone, whose whining house pet was a dog-sized dinosaur named Dino. For sixty years, most Hollywood "caveman" films belonged to a sleazy genre—curvy starlets wearing scanty skins (*One Million B.C.*, 1940) and monosyllabic musclemen beaning each other with papier-mâché boulders on backlot sets. Sometimes rubber dinosaurs terrorized the tribe, as in the Ringo Starr comedy *Caveman* (1981).

Then, in 1982, director Jean-Jacques Annaud released *Quest for Fire*, a conscientious high budget film about early man—without dinosaurs— against which all future efforts will be measured. Annaud showed four distinct types of hominids coexisting in the same time and place. Some were more Neanderthal-like; others were modern humans; and one tribe were fierce cannibals who resembled *Homo erectus*.

Shot in wilderness locations in Canada, East Africa, and the Scots highlands, the bands of humans seem especially small and vulnerable against the vast cinemascope landscapes. Filmed just before the era of computer-generated imagery, Annaud had Asian elephants laboriously "made up" with hair and tusks to resemble wooly mammoths. He also took the trouble to commission zoologist Desmond Morris (author of *The Naked Ape*, 1967) and novelist-linguist Anthony Burgess (author of *A Clockwork Orange*, 1962) to devise a "primitive" language, which combined words, gestures, and primate communication signals.

Interestingly, the film's basic premise—that several different hominids existed at the same times and places—turned out to be almost two decades ahead of anthropology. At the time, in the 1980s, scientists believed that only one hominid at a time could be successful; today, the consensus more closely resembles the scenario in Annaud's film.

Jay Matternes, one of today's preeminent paleoartists, has incorporated in his paintings some present anthropological preoccupations. In several of his works, man the hunter is now aided by woman the gatherer, and females are shown not merely as trophies to be clubbed and dragged off by the hair, but as essential partners in the community's survival. Matternes' magnificent murals of australopithecines and early primates were influenced by his knowledge of recent field studies of primate behavior. In the 1990s, curator Ian Tattersall commissioned two large Matternes paintings for the American Museum's Hall of Human Biology and Evolution. His mural of the early primates will continue to be displayed in the new Hall of Human Origins.

Matternes, like Charles Knight, often begins by creating sculptures as references for his paintings. "Making a preliminary sculpture even a quick one, to study light and shadow is a device frequently used by artists, and I have used it often," he writes. Knight used to take his sculptures of dinosaurs, mastodons, and cavemen to the roof of his New York City studio at different times of day to observe where the shadows fell. Viktor Deak also makes sculptural busts of his ancient hominids first, then photographs them—and finally retouches the photos digitally on a computer.

Like Knight, Matternes also had performed many laborious anatomical dissections. His classic reconstructions of *Paranthropus robustus*, *Australopithecus afarensis* ("Lucy"), and the Shanidar I Neanderthal, he wrote, were produced "by the sequential soft-tissue buildup method, aided by my dissection data of a number of anthropoid apes, which I had performed at the Smithsonian since 1966. Also, I dissected many human heads at Georgetown University Medical School, which provided much helpful information that was not available in the literature."

All the best paleoartists, including Mascré, Knight, Gurche, Matternes, the late English painter Maurice Wilson, and the superb English sculptor John Holmes, have worked closely with anthropologists and anatomists. Viktor Deak's scientific mentor, Gary Sawyer, had studied medical technology, anthropology, and prehistoric archeology, and also had dissected facial muscles of apes and human cadavers.

As a boy, Sawyer had been fascinated by the American Museum's famous bronze busts of Piltdown Man, Neanderthal, Java Man, and Cro-Magnon, which were sculpted by James Howard McGregor (1872–1955) and published in Henry Fairfield Osborn's 1915 classic textbook on prehistory, *Men of the Old Stone Age*. McGregor was a professor of zoology at Columbia University who undertook the reconstructions of prehistoric faces at the urging of Osborn, who used them, along with those of Mascré, to illustrate his books on fossil man. In 1931, McGregor went to Africa on a joint expedition of the museum and Columbia University, which brought back the most extensive scientific data on the anatomy of the great apes ever gathered up to that time. Seventy years later, Gary Sawyer sought out McGregor's busts in the museum's storage, and keeps the originals on view in his laboratory, where they provide daily inspiration.

In 1995, curator Ian Tattersall asked Sawyer to do a new reconstruction of the Beijing *Homo pekinensis* skull from casts of fragments. After his updated revision was widely accepted by anthropologists, Sawyer then made a soft tissue restoration over the restored skull. Next, he went on to tackle Neanderthal. After obtaining bone casts from several Neanderthal partial skeletons scattered in the world's museums, in 2002 Sawyer and colleague Blaine Maley completed the first reconstruction of a complete Neanderthal skeleton, which will now have a permanent place in the American Museum's new Hall of Human Origins.

In Darwin's day, people asked, "Where is the missing link?" Today, as previously unknown varieties of human and near-humans continue to be identified, keeping up with the pace of discovery is a continual challenge. There are so many "missing links" that paleoanthropologists don't know what to do with them all. In a departure from what anthropologists believed a decade ago, the new evidence indicates that several kinds of humans lived on Earth at the same time and in some of the same places. Even as Sawyer and Deak were working on this book, fossils of three remarkable new early hominids came to light: the seven-million-year old Tchad skull (*Sahelanthropus tchadensis*), the "Hobbits" of Flores Island (*Homo floresiensis*), and most recently the infant *Australopithecus afarensis* skeleton ("The Dikika Child"), all of which they duly included.

New resins, rubbers, and plastics give paleoanthropologists finer tools with which to achieve accuracy of form. In addition, they can use medical computer techniques for sectioning or resoring distorted fossil skulls. They can consult films, videos, and long-term studies of primate behavior and human hunter-gatherers that were not available to Knight and Burian

Of course, texture and color of hair and skin are still matters of interpretation, although work with ancient DNA may eventually shed light on these areas, too. The best artist-scientist teams try to keep their imaginations in check, and treat the emerging likeness of a prehistoric face as a puzzle to be solved, according to strict rules of the game. As Gurche puts it, referring to an eight-million-year-old fossil ape discovered in Greece in 1990:

> The final form of the animal is often a surprise—I try not to let any preconceptions guide me. I didn't expect *Ouranopithecus* to look as gorilla-like as it does, for example, but when I followed the process of laboriously applying the technique I developed from my great-ape dissections, that's just the way it came out.

A paleoartist seeks more than a good likeness; he or she seeks to depict behavior, culture, and emotional expression. Writing during the First World War era in Belgium, the anthropologist Louis Rutot emphasized that point about the reconstructions he and sculptor Aimé Mascré had created:

> The first stage of the reconstruction is very gratifying to the scientist, but frustrating to the artist, as he is constantly nudged to get the exact conformations and measurements correct. It is an intimate collaboration....Only when the science is satisfactory, does the aesthetic come into play. Faults are corrected, contrary elements harmonized and at last the face appears—exact but still inert. At that point the fingers of the artist impart thought, movement, and life.

We humans seem incapable of gazing, Hamlet-like, at a bit of skull or jawbone without trying to conjure up an image of how its owner appeared in life—and how similar or different its face appeared from our own. Some stately European homes of rich or royal families contain impressive galleries of ancestral portraits that go back ten or twenty generations, but most of us are fortunate if we have a faded photo of our great-grandparents. And yet, in each generation, a few talented anthropologists, anatomists, and paleoartists—eternally optimistic—will combine their skills in an attempt to show us what our relatives looked like a hundred thousand and more generations ago. Straining to see through time, by yoking science and art to a common purpose, they move forward by looking backwards at the long, winding path we have traveled.

FURTHER READING

Alexander, R. McNeill. 2005. *Human Bones: A Scientific and Pictorial Investigation.* New York: Pi Press.

Berman, Judith. 1999. "Bad Hair Days in the Paleolithic," *American Anthropologist* January, vol. 101, 2:288-304.

Day, Michael. 1986. *Guide to Fossil Man* (4th ed.). Chicago: University of Chicago Press.

Goldfinger, Eliot. 1991. *Human Anatomy for Artists: The Elements of Form.* New York: Oxford University Press.

Gould, Stephen Jay. 1996. "A Lesson from the Old Masters." *Natural History Magazine.* August. Reprinted in *Leonardo's Mountain of Clams and the Diet of Worms.* 1998. New York: Harmony Books.

Johanson, Donald C., and Blake Edgar. 2006. *From Lucy to Language* (rev. ed.). New York: Simon and Schuster.

Klein, Richard. 1999. *The Human Career* (2nd ed.). Chicago: University of Chicago Press.

Klein, Richard, and Blake Edgar. 2002. *The Dawn of Human Culture.* New York: John Wiley and Sons.

Lambourne, Lionel. 1977. *Ernest Griset: Fantasies of a Victorian Illustrator.* London: Thames and Hudson.

Mann, Alan. 2003. "Imaging Prehistory: Pictorial Reconstructions of the Way We Were." *American Anthropologist* March, vol. 105, i:139-143.

Milner, Richard. 1993. "New Days for Old Knights: Restorers Rescue a Museum Artist's Mammoth Masterpieces." *Natural History* October:20-24.

_________. 1995. "Portraits of Prehistory: An English Artist Creates State of the Art Sculptures of Our Remote Ancestors." *Natural History* December:44-48.

Moser, Stephanie. 1998. *Ancestral Images: The Iconography of Human Origins.* Ithaca: Cornell University Press,

Rainger, Ronald. 1997. *Agenda for Antiquity: Henry Fairfield Osborn and Vertebrate Paleontology at the American Museum of Natural History, 1890-1935.* Tuscaloosa: University of Alabama Press.

Sarmiento, Esteban. 1987. "The Phylogenetic Position of *Oreopithecus* and Its Significance in the Origin of Hominoidea." *American Museum novitiates* no. 3091.

_________. 1998. "Generalized Quadrupeds, Committed Bipeds, and the Shift to Open Habitats: An Evolutionary Model of Hominid Divergence. *American Museum novitiates* no. 3250.

Sarmiento, Esteban, Eric Stiner, and Ken Mowbray. 2002. "Morphology-based Systematics (MBS) and Problems with Fossil Hominid Systematics." *The Anatomical Record* 269, 1:50-66.

Sawyer, G.J., and Blaine Maley. 2003. "Neanderthal Reconstructed." *The Anatomical Record,* 2838:23-31.

Sawyer, G.J., and Ian Tattersall. 1995. "A New Reconstruction of the Skull of *Homo erectus* from Zhoukoudian, China." *Proceedings of the International Conference of Human Paleontology* ORCE.

Tattersall, Ian. 1995. *The Fossil Trail: How We Think We Know What We Know About Human Evolution.* New York: Oxford University Press.

_________. 1995. *The Last Neanderthal: The Rise, Success, and Mysterious Extinction of Our Closest Human Relatives.* New York: Macmillan.

Tattersall, Ian, Ralph Holloway, Douglas Broadfield, Michael Yuan, and Jeffrey L. Schwartz. 2005. *The Human Fossil Record* (4 vols.). New York: John Wiley and Sons.

Tattersall, Ian, and G.J. Sawyer. 1996. "The Skull of '*Sinanthropus*' from Zhoukoudian, China: A New Reconstruction." *Journal of Human Evolution,* 31:311-314.

Tattersall Ian , and Jeffrey L. Schwartz. 2000. *Extinct Humans.* New York: Westview Press.

ACKNOWLEDGMENTS

G.J. Sawyer thanks the late James Howard McGregor, Professor of Anatomy, Columbia University, for inspiring me to seek out the secrets of restoring "life from the past"; Ian Tattersall for having faith in the work I do, and for his guidance of and support for the Fossil Hominid Forensic and Reconstruction Team; my medical school mentor, pathologist Eugene Tschekunow, whose teaching and guidance would make all the difference in my career; Philip Tobias for his generous support, consultation, and scientific advice in the creation of many of these reconstructions; Eliot Goldfinger for his hands-on critique of many of the soft tissue reconstruction phases; Donald Johanson for his warm friendship and support; Charles Hilton for his advice in the creation of the first Neanderthal skeleton; Mario Chech of the Museum of Man, Paris, who generously supplied the "missing links" to the Neanderthal skeleton; Robert Franciscus for his kind assistance in the initial research on the Neanderthal skeleton; Henry and Deborah Galiano of Maxilla & Mandible for donation of their lab facilities in the creation of the Neanderthal skeleton; Blaine Maley, who toiled the long hours with me in creating the Neanderthal skeleton and the Lucy skull; Jay Matterness for his dynamic anatomical reconstructions that charged my imagination; Bob Ziering for his untiring participation as a member of The Fossil Hominid Forensic and Research Team; Samuel Marquez for his generous support and anatomical advice; Patience Freeman for all her help and advice on the reconstruction project; Mike Peter Smith with whom has been a pure joy of intellectual glee to work; Aaron Diskin for his support as a member of our Forensic Reconstruction Team; Barry Wizoreck for being a great team player and valuable member of the Fossil Hominid Forensic and Reconstruction Team; Eric Stiner, a great down-to-earth user friendly guy, and who upon looking at the Neanderthal skeleton for the first time stated: "Awesome !"; Ken Mowbray for his computer assistance, and work on the Neanderthal skeleton; and Will Harcourt-Smith for being a truly jolly nice chap, and for all his kind critiques and advice on the reconstructions.

In addition to those recognized about, Viktor Deak would like to thank my mother and my father—this book is for you; my grandfather for the Burian book that lead me here; Xochitl Gomez; Marika Aranyi for your belief; Nusi Neni for watching over me; the School of Visual Arts; Kim Ablondi, Marshall Airisman, and Bernice M. Nicolari for your wisdom; John Gurche, Jay Matternes, the Leakey Family, and Charles Knight for inspiring the search for the truth; Charles Hearn, Alexis Rockman, Reuben Negron; Jeremy and Dave Bronson, Ben Trinh, and Matt Morris for your great friendships; Barry Wizoreck, for your help in Aladin's Cave; Arfija Imotovski—visualize to materialize; The SECRET; Luis Ortiz; Tonia, Paul, Augie, and Perry Barringer, and Ian Falconer; Joe Horton, Tom Dunn, Joe Demarco, the Crasilli family, Dylan Griffin, and Denis Fazekas for support; Regina and Americo Carchia; Carla, Andy, Angela, J.R. and S unit for your guidance; Tizoc Gomez, John Deak, Charles Hearn, Samantha Coleman, and all the models involved; and the American Museum of Natural History for the opportunity to be apart of your legacy—your halls are the birthplace of dreams.

Esteban Sarmiento thanks Doug Hajicek, John deVos, Phillip Tobias, and Chris Donahue.

Richard Milner thanks antiquarian bookseller David Bergman of New York City for illustrations from rare books. Photo of the wodwose statue on St. Mary's Church (Mendlesham, UK) was kindly provided by John Telford-Taylor. John Gurche, Jay Matternes, Ian Tattersall, Vittorio Maestro, Judith Berman, Peter Brown, Lindsay Fulcher, Elizabeth Donohue, and film historian David Shepherd For helpful comments. Darwin scholar Randal Keynes generously provided expert guidance. Special thanks to the Bromley Museum in Orpington, UK, and its curator, Adrian Green; and Gerry Ohrstrom, Susan Wilder, Norman Shaifer, and the late C. R. Tripp for continuing support and encouragement.

All the contributors to ***The Last Human*** thank Cathleen Elliott for her untiring work to create a truly elegant book design, and Jean Black, Yale University Press editor, for having faith in the concept and the overall success this kind of book promises.

INDEX